# Aesthetic Sustainability

Why do we readily dispose of some things, whereas we keep and maintain others for years, despite their obvious wear and tear? Can a greater understanding of aesthetic value lead to a more strategic and sustainable approach to product design? *Aesthetic Sustainability: Product Design and Sustainable Usage* offers guidelines for ways to reduce, rethink, and reform consumption. Its focus on aesthetics adds a new dimension to the creation, as well as the consumption, of sustainable products. The chapters offer innovative ways of working with expressional durability in the design process.

*Aesthetic Sustainability: Product Design and Sustainable Usage* is related to emotional durability in the sense that the focus is on the psychological and sensuous bond between subject and object. But the subject–object connection is based on more than emotions: aesthetically sustainable objects continuously add nourishment to human life. This book explores the difference between sentimental value and aesthetic value, and it offers suggestions for operational approaches that can be implemented in the design process to increase aesthetic sustainability. This book also offers a thorough presentation of aesthetics, focusing on the correlation between the philosophical approach to the aesthetic experience and the durable design experience.

The book is of interest to students and scholars working in the fields of design, arts, the humanities and social sciences; additionally, it will speak to designers and other professionals with an interest in sustainability and aesthetic value.

**Kristine H. Harper** is an Associate Professor in Sustainable Fashion at Copenhagen School of Design and Technology. Her research focuses on sustainable product design, philosophical aesthetics, aesthetic nourishment, the durable subject–object bond, and Designer Social Responsibility. She has written a wide range of articles on sustainable design and durable aesthetics.

## Routledge Studies in Sustainability

https://www.routledge.com/Routledge-Studies-in-Sustainability/book-series/RSSTY

**Rethinking the Green State**
Environmental governance towards climate and sustainability transitions
*Edited by Karin Bäckstrand and Annica Kronsell*

**Energy and Transport in Green Transition**
Perspectives on Ecomodernity
*Edited by Atle Midttun and Nina Witoszek*

**Human Rights and Sustainability**
Moral responsibilities for the future
*Edited by Gerhard Bos and Marcus Düwell*

**Sustainability and the Art of Long Term Thinking**
*Edited by Bernd Klauer, Reiner Manstetten, Thomas Petersen and Johannes Schiller*

**Cities and Sustainability**
A new approach
*Daniel Hoornweg*

**Transdisciplinary Research and Practice for Sustainability Outcomes**
*Edited by Dena Fam, Jane Palmer, Chris Riedy and Cynthia Mitchell*

**A Political Economy of Attention, Mindfulness and Consumerism**
Reclaiming the Mindful Commons
*Peter Doran*

**Sustainable Communities and Green Lifestyles**
Consumption and Environmentalism
*Tendai Chitewere*

**Aesthetic Sustainability**
Product Design and Sustainable Usage
*Kristine H. Harper*

# Aesthetic Sustainability

Product Design and
Sustainable Usage

**Kristine H. Harper**

Routledge
Taylor & Francis Group
LONDON AND NEW YORK

earthscan
from Routledge

First published 2018 by Routledge

2 Park Square, Milton Park, Abingdon, Oxfordshire OX14 4RN
711 Third Avenue, New York, NY 10017

*Routledge is an imprint of the Taylor & Francis Group, an informa business*

First issued in paperback 2018

*British Library Cataloguing-in-Publication Data*
A catalogue record for this book is available from the British Library

*Library of Congress Cataloging-in-Publication Data*
A catalog record for this book has been requested

ISBN: 978-1-138-72861-5 (hbk)
ISBN: 978-1-138-36918-4 (pbk)

Typeset in Times New Roman
by Keystroke, Neville Lodge, Tettenhall, Wolverhampton

Translated from Danish by Rasmus Rahbek Simonsen

Cover design:
Concept and idea: Judith ter Haar, intuitive forecaster and curator
Visualisation: Janneke van Rooijen

"It is the rhythm of movements, the intuitive hand and the collected waste of the sea that made this visual work into a jewellery memory."

Drawings by Kristine Hornshøj Harper.

"Kristine Harper has written a great book on a relevant topic we all ought to contemplate and react to. She is well informed of the history of aesthetic ideas and has managed to transform her knowledge innovatively into an independent, well-written and inspiring research contribution that can also be read by a broader audience."

*Professor Dorthe Jørgensen, Philosophy and*
*History of Ideas, Aarhus University, Denmark*

"As the window for action against irreversible climate changes is narrowing, Harper offers timely and practical advice on how, as designers and consumers, we can take responsibility for creating a sustainable future. Though informed by a deep understanding of the complexities of aesthetics and design, her book is highly accessible."

*Per Galle, Associate Professor, The Royal Danish Academy*
*of Fine Arts, School of Design*

"With *Aesthetic Sustainability* Kristine Harper shines light on the role of aesthetics and how we as humans emotionally connect with the objects that surround us in everyday life, and through that, she manages to humanize the concept of sustainability."

*Hanka van der Voet, Head of MA Fashion Strategy*
*at the ArtEZ University of the Arts, The Netherlands*

"Most books on sustainable design focus on the effects of design on the environment mainly in terms of recycling, reuse and repair. Kristine Harper's book introduces the additional and undervalued importance of the aesthetic qualities of sustainable design. She argues that aesthetics are key to creating sustainability that is lasting, due to the added emotional values that both appeal to and nourish the user. A factor most sustainable designers have ignored. The book discusses the interaction and relationship between the three concepts, design, sustainability and aesthetics in depth, thereby giving the reader an (almost) how-to guide to producing aesthetic, sustainable and durable design. A book to be recommended for both professional practitioners as well as students of design."

*Karen Blincoe, Director, Chora Connection*

"The book addresses a highly relevant subject from an unexpected angle. This broadens our perspectives on sustainability to be about more than reusing and recycling but also about providing aesthetic experiences. In particular, the model for an aesthetic strategy provides a useful tool, as it brings forth relevant concepts for exploring and reflecting on choices of expression and how such choices might affect the perception of a design product."

*Per Liljenberg Halstrøm, PhD, postdoc and lecturer at Copenhagen*
*School of Design and Technology and The Royal Danish*
*Academy of Fine Arts, School of Design*

For Sarid, Marius and Severin

# Contents

# Introduction

When my grandmothers died, I inherited a lot of different things: dinner plates, cups, vases, porcelain figurines, tablecloths, knitted cardigans and shawls, jewelry, a toaster, and even a laundry basket. Most of the things I inherited are decorative artifacts; the functional objects such as the toaster and the laundry basket simply underline the fact that sustainability, and not just buying new things for the sake of it, used to be an integral part of our culture; why buy a new laundry basket, if the old one still works? My grandmothers were from a time when placing porcelain figures—such as playing bears, chubby babies with tummy aches or headaches, or majestic cats—around the house indicated good taste, and when a woman would carry jewelry to signal status and wealth. I am a part of a Zeitgeist and a culture where decorative elements in a home are generally limited to potted plants on the window sill or the floor and abstract paintings on the walls, and, personally, I have never cared much for jewelry; as a matter of fact, my favorite jewelry selection is limited to a simple bead necklace from India and plastic bead bracelets made by my sons. Despite this, I cherish the items I have inherited from my grandmothers. Both of the ladies meant a great deal to me, and I have always felt emotionally connected to them. So I have kept all the vases, cups, tablecloths, porcelain figures, and jewelry, along with all the knitwear. However, most of these things never see the light of day; they are kept in my storage room, nicely packed away. Why? Because of their aesthetics. I keep them because of the emotional bond between my grandmothers and me, but I don't feel like decorating my home with them, and I don't feel like wearing them. They just don't *work* for me, *aesthetically*. They are either too kitschy or too charged with the trends or values from a bygone era that have no meaning to me and don't line up with my values or preferences. Yet some of the things are different; they have actually found a place in my home and are worn quite regularly. They have the rare quality of being *charged* with both emotional value *and* aesthetic durability. Their expressions appeal to me—and *nourish* me. They remind me of my grandmothers—and because of the memories they put a smile on my face whenever I look at them or wear them—but they also *work* aesthetically, despite the fact that their expression is rooted in the past.

My intention with this book is to unwrap the many facets of aesthetic sustainability or the aesthetically durable object. What makes a design object's expression or *look* durable? Or, why do we dispose of some things before their use has expired, while others are kept and repaired time and again, despite their obvious wear and tear—or despite the fact that their *look* belongs to a different time or era? Is it possible to define universal aesthetic factors that work for all (or most) human beings?

Sustainable design and production solutions often focus on creating products that can be recycled efficiently and without harming the environment. This book's thesis is that the most significant and fundamental way to transform the world of design is by creating solid, well-crafted, and enduring objects. Sustainable design is also about influencing consumers to purchase fewer but better things. Products are considered sustainable when they can be adapted to different situations and user needs in a functional but also aesthetic and expressive way.

The aesthetic perspective challenges the idea that sustainable design is about breaking down and rebuilding. Instead of focusing on how to *convince* consumers to buy certain things, empowering people to make ethical and aesthetic choices to limit overconsumption is the expressed goal of working with aesthetic sustainability. From an aesthetic point of view, sustainable design should aim to create enduring objects that can be repaired, upgraded, and/or reused. At the same time, sustainable objects must be attractive enough that users will want to keep repairing and reusing them; they must meet the human need for *aesthetic nourishment*.

It is of course important for the environment that designers use sustainable, non-composite materials, and that they seek to minimize transportation time, thereby reducing $CO_2$ emissions. Furthermore, consumers should be encouraged to wash their clothes less and avoid using tumble dryers; products should be easy to repair and upgrade; and strategies that include minimal waste (zero-waste strategies, etc.) should be emphasized during the design process. However, the most sustainable solution is minimizing, or even stopping, overconsumption by educating consumers, via design, to buy fewer but better products—durable products, in terms of their quality *and* aesthetic value. These are objects that consumers will *want* to look after, repair, and, ultimately, pass on. Rather than discovering methods for reusing the leftover materials of broken objects, working with aesthetic sustainability means striving to create objects that will last for a long time and can be used until they have exceeded their natural utility.

The purpose of this book is to establish guidelines for reducing, rethinking, and reforming patterns of consumption with the overall goal of stopping overconsumption through design. These guidelines are designed to counter the self-contained consumption and throw-away mentality that, despite the increased focus on sustainability, continues to dominate our late-modern society. The way forward is to create durable objects that can be repaired and updated by anybody once they finally break or cease to function according to their original purpose. Such objects won't appear obsolete or outdated after a single season but will continue to exude an expressive and aesthetic force for decades to come.

On a basic level, the purpose of design is to create products that contain a sense of value for the user. But how to define "value"? What does it mean to create *aesthetically* sustainable value? And (how) can aesthetic value add *sustainability* to an object? What does "sustainability" mean in the context of objects, things? Is it possible to produce specific "guidelines" for the potential creation of aesthetic and sustainable objects?

My immediate hypothesis—which forms the foundation for the aesthetic analysis of sustainability—suggests that the most durable expression will be easy to decode and will appear balanced and in proportion to its message. This kind of expression contains a certain degree of "neutrality" and minimalism. "Neutrality" is here understood to refer to harmonious objects that may appear in many different contexts and appeal to a broad range of tastes—they comprise a kind of expressive or common universality. Furthermore, aesthetic sustainability is joined to expressions producing a form of pleasure that is founded on the expectation that basic rules regarding symmetry, harmonious color schemes, and design materials are maintained. Sustainable designs must also preserve a certain degree of aesthetic flexibility to ensure their contextual adaptability.

I am, however, also fascinated by an opposite, yet related, question: can a sustainable expression be so complex and challenging that it demands a sustained interest in exploring it (for an extended period of time)? Is the most aesthetically sustainable object a thing of such elevated complexity that the user is immediately (and time again) challenged and forced to consider its provenance in relation to the surrounding world? Perhaps the pleasure of sustainable objects lies in their ability to disrupt the user's comfort zone as, combined with the use of unexpected materials, color schemes bleed and asymmetrical shapes confront our powers of perception and conception. Understood in this way, sustainable design objects possess a multifunctional dimension (in either a practical or aesthetic sense) that enhances their aesthetic flexibility.

Neither approach necessarily cancels out the other. The point is rather that each forms a different, overall strategy for analyzing and working with aesthetic sustainability. The first approach is here defined as, "the Pleasure of the Familiar," and the second as, "the Pleasure of the Unfamiliar." In relation to those things I inherited from my grandmothers that have the rare quality of being *charged* with both emotional value and aesthetic durability, it is significant to note that they share the following exact characteristics: they either exude a pleasurable "neutrality" that makes it possible to combine them with other objects or to wear them in many different situations—*or* they possess a complexity or "quirkiness" that "prods" me, every time I look at them or use them.

## The concept of durability

As outlined in the previous section, the aesthetic experience can be determined according to two kinds of pleasure: the Pleasure of the Familiar and the Pleasure of the Unfamiliar. Both forms of pleasure are enchanting and satisfying, albeit in vastly different ways. The fundamental difference between the two contains a historical reference to the distinction between the beautiful and the sublime—or, rather, to the historical distinction between the beautiful and the sublime aesthetic experience. The following chapters will examine the beautiful and the sublime in relation to the concrete design experience as well as to aesthetic sustainability.

However, to begin, it is important to further define "sustainability" and what I have been referring to as "durable expression."

"Durability" is central to the notion of sustainability and to the development of sustainable design objects and concepts:

- Durability is associated with the use of sustainable, enduring materials, or materials that age gracefully.
- Durability, as a concept, can refer to materials that make it easy to repair or to upcycle design objects.
- Durability is often connected with design solutions that can be updated continuously by means of technology or replaceable elements that safeguard against their obsolescence.
- Durability, finally, can refer to functionality and flexibility.

The focal point for this book is aesthetic sustainability, and my inquiry will attempt to answer the following questions: why do we throw out certain objects and clothing long before their use has expired, whereas we cling to other kinds of objects that we keep and maintain for years, or even for life, mourning any wear and tear they might endure? How is it possible to create this kind of "timelessness" and continual attraction when it comes to things?

Aside from their "timelessness," aesthetically sustainable objects contain an immediate aesthetic appeal. What happens in those moments when we are confronted by an aesthetically appealing object that enunciates the core of who we are as people? Often, we experience an immediate connection to objects, similar to the irrational feelings that characterize the first moments of falling in love with someone. The experience is one of correspondence; the object appears contiguous with our own personal values, sustaining and somehow even expressing them. This kind of emotional experience can lead to an enduring bond, fostered by love, respect, and the desire to maintain and nurture the object in question.

However, can experiences of immediate attraction and the development of enduring bonds be "formalized"? Is it possible to advance general guidelines for what an object has to do, look or feel like, in order to evoke such powerful emotions, which, ostensibly, are subjective and linked to personal experiences and preferences? The two first chapters, "The Pleasure of the Familiar" and "The Pleasure of the Unfamiliar," attempt to do just that.

## Structure and overview

The book revolves around a number of explorations delving into the aesthetic expressions associated with sustainability as well as the different design-strategic considerations regarding the planning or engineering of aesthetic and durable design experiences. Each analytic and strategic consideration presented has to do with sustainability. The analytic elements are partly based on my own observations and partly on the philosophical works of Edmund Burke, Immanuel Kant, Johannes Itten, Wassily Kandinsky, Jean-Francois Lyotard, Roland Barthes, Willy Ørskov, and Dorthe Jørgensen, among others; furthermore, the book draws on the analytical design works, chiefly, by Kate Fletcher, Jonathan Chapman, and Stuart Walker. The analyses and design strategies focus on objects and things rather than on immaterial and conceptual design solutions. Regardless, as will be shown, the book's reflections on design strategies can easily be translated to concept development and storytelling contexts.

This book can be used actively in the creation of aesthetically sustainable design products, as guidelines for working with the production of aesthetic value and effects will be presented. Moreover, the book contains directions for design analysis and provides insights concerning philosophical aesthetics as well as the intersections between the traditions of aesthetics, design, and sustainability.

The book is divided into the following sections:

- Chapters 1 and 2 comprise the book's theoretical and philosophical foundation. Here, the focus is on the beautiful and the sublime. The chapters translate these philosophical and aesthetic categories to "the Pleasure of the Familiar" and "the Pleasure of the Unfamiliar," both of which can be used in analyzing and planning the aesthetic experience that a design object may elicit. The purpose in the two first chapters is partly to present a distinction between what it means to decode and, conversely, to simply detect an object.

The remaining chapters contain a number of perspectives on aesthetic sustainability. Additionally, they provide different suggestions for creating strategic guidelines that can be used when working with aesthetically sustainable products:

- Chapter 3, "The expression of flexible aesthetics," focuses on the concept of "beauty"—specifically, whether beauty is fleeting and changeable or permanent in its expression. The chapter also deals with how to create broadly appealing design objects. Furthermore, methods are presented for Zeitgeist analysis.
- Chapter 4, "Designing the temporal object," deals with how designers can work *time* into an object and thereby increase its emotional and aesthetic value. This could be done through narrative means connected to the design process, by using recyclable materials, or by creating an illusion of ageing or decay.
- Chapter 5, "The magical thing," concentrates on the aura or magic that surrounds certain things, as well as the emotional tie we often establish to *magical* things. By analyzing the concept of "thing-magic," the chapter establishes guidelines for how this can be incorporated into the product design.
- Chapter 6, "The value of aesthetic sustainability," focuses on how best to communicate the message of sustainability and aesthetics. Unless consumers have access to the kind of thinking and time that have formed the aesthetically sustainable object—or are able to experience the sensuous pleasure that characterizes the good design experience—it might be difficult to convince them that the object is in fact durable, and that, consequently, its expression

and quality will continue to elicit a sense of aesthetic pleasure for years
to come.

- Chapter 7, "Aesthetic strategy," offers a synthesis of the book's different
considerations. The chapter presents a model that can be used for working
strategically with aesthetics in the design process as well as for planning
the recipient's aesthetic experience. The aesthetic strategy model integrates the
principles of the Pleasure of the Familiar and the Pleasure of the Unfamiliar,
constructing a set of opposites to guide the design process. As demonstrated in
the chapter's final section, the model can also effectively be used to analyze
objects and concepts.

# 1    The Pleasure of the Familiar

There is a great deal of satisfaction, or even a great deal of pleasure, in experiencing one's milieu and the objects in it being exactly as expected. This kind of satisfaction is related to the comfortable feeling of security that we experience when facing familiar phenomena and events, and it basically consists of knowing precisely what is expected of us, and how we should act and interact with others. It is a satisfaction linked to order and harmony and particularly to predictability; furthermore, it is closely connected to the human need to structure everyday life and create routines for ourselves.

The Pleasure of the Familiar results from experiencing the artifacts in one's surroundings as being, acting, and working just as expected:

- It can be related to touching the surface of a table, expecting a certain familiar tactile experience—coolness, smoothness, or hardness—and getting exactly that experience.
- It can consist in putting on a dress and experiencing that the fit is exactly as expected, and that it wraps the body in precisely the way that one anticipated: the feeling, or touch experience, matches the look of the fabric.
- It can refer to the experience of holding a jacket in a shop and immediately being able to decode or understand how to put it on, how to open and close it.
- It can be associated with the experience of a blanket or a shawl having a well-balanced color combination that creates a harmonic expression of comfort and safety.
- It can also be described as the pleasure of pulling a chair out from under a table and experiencing exactly the anticipated degree of weight as well as the expected surface touch experience—meaning the tactile experience of the material being in accordance with the visual experience.
- It can be the comfortable and pleasant experience of being immediately able to use a kitchen utensil that has not previously been held or used—a potato peeler, a corkscrew, a coffeepot or a spoon—simply because its shape and functionality can be understood based on mental references to other similar objects. This sort of comfortable design experience is also related to the fact that the way the object should be used is *inherent* to its shape; our hands just intuitively know what to do.

- It can, more abstractly, be described as the pleasure of seeing how the material effortlessly or naturally fits the shape of the object. That the material, in other words, doesn't possess a great deal of *inertia* in relation to the shape.[1]
- It can be the satisfying experience of entering a café never previously visited and realizing that it is designed and structured in exactly the same way as other, familiar cafés. Due to the familiar layout, one will know how to act, for example, in finding a table, ordering at the counter, paying for the order, and returning to one's seat.
- Or, it can be related to the pleasurable experience of finding one's way effortlessly through a city or a building by following easily decodable pictograms or easily understandable signs.

We all need comfortable habitual or habit-generating experiences like the above. Without experiences like these, our surroundings and milieu would feel like one big unmanageable chaos, which would require a lot of energy and strenuous activity to navigate. We need to feel that we are, at least to some extent, in control of our surroundings, that we can grasp and understand the artifacts and the space structures that surround us. And we need to know the "ground rules" of our everyday milieux— a need that is rather physical or connected to bodily experience. Furthermore, we need to feel safe and that we belong (to other people and spaces); we also need to know how to act in the "right" way socially and culturally as well as in relation to the objects and artifacts that surround us. If the objects we come across in our daily lives work as expected or provide us with the anticipated sensuous impressions, we are able to meet our human need to feel in control; to feel safe; and to act within comfortable, familiar settings.

Aesthetically, the Pleasure of the Familiar can be linked to the concept of the beautiful. The Pleasure of the Familiar, after all, is linked to the human need and ability to apprehend and use common, everyday objects, as this boosts our physical comfort level. Therefore, in order to define and analyze the durable aesthetic experience concerning the Pleasure of the Familiar, I will begin by introducing the term *the beautiful* in its historico-aesthetic context and in relation to the aesthetic experience.

## The beautiful

When working with aesthetics and the aesthetic experience, there are two fundamental and important terms to consider: the beautiful and the sublime. These two terms, which in many ways represent two fundamentally different sides of the aesthetic experience, have been made subject to numerous philosophical theses and art historical discussions. In the following section, I will draw on a few of these, since they have influenced my interpretation of the durable expression and aesthetic sustainability.

The historical division between the beautiful and the sublime indicates that an aesthetic experience is not necessarily linked to beauty, but can also be induced by the unpleasant, unbalanced, distorted, or even hideous. This could include a dilapidated old house or a frightening demonic figure in a gothic church.

The most significant differences between the beautiful and the sublime can be outlined like so:

| BEAUTIFUL | SUBLIME |
| --- | --- |
| Symmetry | Asymmetry |
| Comfortable | Uncomfortable |
| Order | Chaos |
| Predictability | Unpredictability |
| Demarcation | Limitlessness |
| Shape | Shapelessness |
| Balance | Unbalanced, distorted |

Briefly put, the beautiful can be defined as a mode of expression that complies with the aesthetic ground principles concerning, for example, color harmonies and composition. As the counterpoint to the beautiful, the sublime characterizes phenomena or objects that provide receivers with a kind of aesthetic pleasure that doesn't match the "classical" concept of beauty, or that disrupt the universal ground rules of aesthetics. I will return to the concept of the sublime in Chapter 2, but first we must understand and analyze the concept of beauty as it relates to aesthetic sustainability and design.

The beautiful is mainly linked to *shape* or to proportional, harmonic objects that provide the receiver or viewer with immediate pleasure. The connection between the beautiful and shape, or proportion and balance, is rooted in ancient times. Aristotle (384–322 BC) describes in his *Metaphysics* how the Pythagoreans (from the 6th century BC) viewed the world's manifestations as mathematically structured and determined by numeric relations. To the Pythagoreans, beauty was identical to order and thereby not only linked to the human experience of the world, but rather to something absolute, something unchangeable and universal. Beauty was seen as the sum of the world's harmonious, symmetrical, proportional forms (Jørgensen 2008: 29).

To Plato (c. 428–348 BC), a physical object can be considered beautiful if it clearly expresses the *form* or *idea* that gave birth to it. A beautiful chair, according to this way of thinking, is a chair that is clearly recognizable as being a chair and that is *good* at being a chair. There is thus a kind of *precision* to beauty (Böhme 2010: 24). Beauty is precise and unambiguous in its expression. Beautiful physical objects are clearly expressed and decoded as what they are, at the same time as they are good at being what they are. Such a viewpoint contains the germ of a functionalist approach to thinking aesthetics. Functionalism, a term for defining the style and historical context of early 20th century design and architecture, is dominated by simplicity and objectivity, understood as form being subservient to function. "Form follows function" is a well-known adage ascribed to functionalism, which aimed to cleanse form of anything but the absolute most necessary elements. The famous phrase was uttered by the American architect Louis Sullivan (1856–1924), and it was in direct opposition to the organic decorative idiom of the previous art-nouveau period. The Bauhaus architect and furniture designer Marcel Breuer (1902–1981) called his chairs "sitting machines" and hereby gestured to the ancient idea that a chair is beautiful if it is good at being what it is. A chair becomes a sitting machine insofar as it is good to sit in. The form and expression of an object are thus inferior to its function. This is the functionalist definition of beauty.

In part, then, beauty is about functionality. Or, at least, about a precise, unambiguous minimalist idiom that is easy to decode or "take in."

In the Platonic dialogue *Greater Hippias* (c. 390 BC) Socrates and Hippias are searching for a definition of beauty, and as part of this search, they try to determine whether a spoon made from gold is more beautiful than one made from fig wood (Plato 1997: 908). Socrates feigns uncertainty. Of course, a golden spoon is finer (and thus more attractive) than its wooden counterpart, but it is harder to handle when eating soup. In the final analysis, the wooden spoon is more beautiful since it is better at being what it is (that is, a spoon); it is more functional, useful. In Plato, beauty is linked to the *good*. In this way, for an object to be considered beautiful and, hence, durable, the material must, crucially, follow form. In Chapter 4, I will go into more detail about the experience of materials.

Plato's way of correlating beauty with the good (or the functional) is in contradiction with my initial hypothesis that aesthetic appeal is foundational to aesthetic durability: a kind of senseless attraction or emotional attachment, which is largely irrational, and which is thus not based on an assessment of whether the object in question adequately fulfills its prescribed function. An object's immediate power of attraction can render inoperative any critical assessment of its functional qualities; in turn, once momentary fascination fades, this can lead to a sense of frustration.

On the one hand, it can seem right that an object should be good at being what it is (it is after all pleasurable to experience and interact with an object that seems to be perfectly aligned with its functional basis of existence). On the other hand, there are objects whose primary function seems to be that of producing aesthetic pleasure for human beings. We would call such objects aesthetically functional. Functionality should thus not be understood solely in terms of usefulness, and consequently, usefulness should not be the only determining factor for assessing an object's durability. In the context of the beautiful, in contrast with the sublime, functionality can therefore be defined as accommodating humanity's preference for the orderly, the proportional, and the well structured.

A preference for proportion, balance, and symmetrical form—which characterizes classic Greek art—entails that the purest geometric forms are considered the most beautiful, as their mathematical proportionality is the most simple. In this approach to form, which is largely based on maintaining proper proportionality, beauty is consummate with the *idea* of beauty, or what characterizes all beautiful objects (Jørgensen 2008: 39). The idea of beauty is the essence of beauty, and the ultimate chair is thus the essence of what a chair is and does, like the spoon to end all spoons.

Based on my ambition to construct concrete guidelines for creating aesthetically sustainable design, I want to determine a common denominator for all beautiful things. The question is, however, if beauty is to be found alone in things, or if it is rather to be found in the interaction between object and subject. As I am primarily interested in exploring the aesthetic experience, I will presume that crucial to determining the essence of beauty is namely the *experience* of things and thus the interaction between subject and object.

Regarding aesthetics, Danish professor of philosophy and the history of ideas, Dorthe Jørgensen writes:

> Beauty isn't just an aspect of the thing we can apprehend and refer to as beautiful. Beauty isn't merely objective; it doesn't appear in the world as a given. But beauty isn't subjective either; it is not simply in the eye of the beholder.

She continues:

> Beauty exists in the meeting point between a potentially beautiful object and a subject who, by virtue of her gaze, makes possible the experience of beauty as such.
>
> (Jørgensen 2012: 35; transl.)

Two central elements appear here: 1) the object of potential beauty and 2) the subject who has the ability to experience beauty. The next two sections will therefore look at the potentially beautiful object to determine whether it is possible to set down universal criteria for judging objective beauty. Additionally, the sections will deal with the experiencing subject and her cultural "baggage" or connotative frame and its importance for potential aesthetic experiences. The question becomes: can we be brought up with or develop the capacity for such experiences?

## Adhering to universal aesthetic principles

When can an object be experienced as being beautiful? Following the previous section, the beautiful aesthetic experience can be defined as the experience of harmonious form and the proportional, the symmetrical, and the demarcated; in other words, beauty is a "smooth" experience where nothing impedes one's ability to apprehend or take in the object or phenomenon one encounters. However, as human beings are all different—with different preferences, tastes, lifestyles, and cultural backgrounds—everybody has a different opinion about what counts as beautiful, exciting, moving, captivating, etc. For this reason, does it even make sense to talk about universal aesthetic principles?

The short answer is that, yes, it does make sense. Of course, personal taste, trends, and the Zeitgeist are all involved when discussing aesthetics and beauty. Nevertheless, there are universal, or common, human preferences concerning which elements make up the most comfortable expression or idiom. People may be different, but we all share certain physiological traits, and likewise, our senses respond to stimuli in much the same way. For that reason, despite our many differences, it might still be possible to construct a set of principles for how shapes, colors, and materials are experienced and internalized by the senses, and for how *a priori* (i.e. before interpretation and tacked-on meanings are applied) which expressions are considered the most balanced and immediately apprehensible.

Throughout history, a number of thinkers have done exactly that. In what follows, I will make reference to some of these in order to theoretically define the comfortable aesthetic expression, which may lead to the Pleasure of the Familiar.

### The need for structure and balance

The German philosopher and psychologist of perception Rudolf Arnheim (1904–2007) considers a natural function of sight to actively select and categorize; for example, oval shapes are spontaneously and immediately categorized as variations on circles. In the effort to order and understand our surroundings, the human sense organs will naturally seek out forms that are easy to recognize and label.

When human beings apply their senses, they seek to structure the world, and this can be described as a *tactile* process that is aimed at sorting out coherent shapes from the many distorted surfaces and forms of the world. As Arnheim puts it:

> In looking at an object, we reach out for it. With an invisible finger we move through the space around us, go out to the distant places where things are found, touch them, catch them, scan their surfaces, trace their borders, explore their texture. Perceiving shapes is an eminently active occupation.
>
> (Arnheim 1974: 43)

In *Art and Visual Perception*, Arnheim lays out a number of concrete principles for spontaneous universal human visual experiences. For example, it may be that "an unbalanced composition looks accidental, transitory, and therefore invalid" (Arnheim 1974: 20), and consequently, designers seeking to satisfy the spontaneous universal human pleasure derived from balanced expressions should work toward creating well-balanced pieces. The human eye is looking for balance and harmony, and it will thus be "repelled" or confused by unbalanced pieces. According to Arnheim—despite the fact that he uses the term "to conceptualize" to describe what happens in the moment when the eye seeks to organize, structure, and take in the physical world—to sense is not an intellectual act. Be that as it may, there are yet certain correspondences between the elementary sensing process and rational deduction.

Arnheim suggests that, as part of the creative process, one can strive to create a greater or lesser degree of consistency or variation, something that is highly relevant to the context of design. Designers can choose to create an expression that is either easy to decode or more complex (I will return to this in Chapter 2). Furthermore, by varying the basic shape of one's design, it is possible to appeal to the recipient's preference for unity and balance, allowing her gaze to wander across and juxtapose different forms and expressions. It is important to note, according to Arnheim, that this preference for order is universal, common to all people regardless of cultural or social "baggage."

As a designer, it is of course possible to challenge the universal and instant-aneous drive to juxtapose and categorize sense impressions in order to understand them. However, since the human gaze is constantly looking to structure and order its surroundings, by accommodating this drive design objects can create a high degree of immediate satisfaction, instant payoff. This immediate kind of satisfac-tion is connected to the Pleasure of the Familiar and to the beautiful aesthetic experience.

According to Danish sculptor and philosopher Willy Ørskov (1920–1990), the universal need for structure is rooted in a quest for accessibility; rhythmic repeti-tion and harmonious sections (which the eye is always hunting for) help eliminate from our surroundings "dread," or disorder and chaos. As Ørskov puts it, "The will to and need for structure is elemental and universal to humanity. It is expressed by a striving for division, order, focus and rhythm" (Ørskov 1987: 88). This process of striving for harmony, order, and rhythmic repetition is universal, common to all human beings. Everybody, regardless of cultural background and stylistic prefer-ences, is always looking to structure his or her surroundings, and for this reason, a mode of expression based on rhythmic repetition, symmetrical structures, and harmony will be the most pleasing to the human eye, in an immediate sense.

The Pleasure of the Familiar is exactly defined by this sense of comfortableness. It is a kind of pleasure that is characterized by accommodating the human need for structure, order, and focus, as well as control and clarity.

Based on a functionalist and constructivist approach to design, instructors at the Bauhaus School[2]—such as Walter Gropius (1883–1969), Wassily Kandinsky (1866–1944), Johannes Itten (1888–1967), Paul Klee (1879–1940), Ludwig Mies van der Rohe (1886–1969), Herbert Bayer (1900–1985), and Marcel Breuer (1902–1981)—attempted to locate a universal idiom that would give form to an international style, a style that would cancel out local differences and that could be decoded and appreciated by all, regardless of cultural background or contemporary preferences. Function was superior to form, just as function should adapt to the needs of the human body.

For the Bauhaus designers, the minimalist, functionalist expression—which they sought, investigated, and applied—was a universal idiom since it, devoid of ornament and symbolism, was able to minimize "misunderstandings" in decoding objects. Moreover, the idea was that a minimalist and "neutral" design object would fit more easily into a number of contexts—or put differently, it would be highly adaptable to its surroundings. This adaptability, and in turn, durability, stemmed from the fact that it didn't create any *visual noise*. The quest for a durable design expression was immanent to the Bauhaus mission, which was focused on creating time- and placeless expressions. The universality of Bauhaus was thus linked to a direct disassociation from time and place.

Moving away from an aesthetic that is tied to a time and place, according to Bauhaus, can minimize the risk of aesthetic decay or outdating, which by definition is attached to cultural preferences for "good taste" that are influenced by the Zeitgeist. In this way, a mode of expression that manages to stay time- and placeless, due to its universality, can be considered beautiful.

## *The pure expression*

The term "minimalistic" has already been used several times in this chapter to describe the kind of expressions that are linked to the Pleasure of the Familiar. Minimalist design is understood to shun ornamentation in favor of "clean," balanced, and quickly decoded expressions. For this reason, I will briefly describe the historical minimalism that contains the essence of the term "minimalist."

Minimalism erupted on the scene in the U.S. of the mid-1960s in the guise of artists such as Donald Judd (1928–1996), Sol Le Witt (1928–2007) and Frank Stella (b. 1936). These minimalist artists sought to cleanse the current idiom, to wipe clean the slate, in order to create a new foundation for art. By reducing art to the basic shapes (circles, triangles, and squares) and to the primary colors (red, yellow, blue, as well as black and white, in keeping with Bauhaus), they were able to create symmetrical pieces consisting of highly homogenous elements.

The visual principles of minimalism are simplification and transparency; there are no complex messages and no abstract or cryptic codes in the form of titles to pieces. The minimalist work of art is open, or free of origins, in the sense that it doesn't refer to any specific phenomenon or any specific event (which would require a pre-existing knowledge or cultural "baggage" to understand). Instead, the point of reference is purely spatial. The work is part of the space, and it thus reaches into the world of the viewer. In this way, the experience becomes exceedingly physical, corporeal. The important thing about the art experience is how it feels to be present with the work and not the associations and stories embedded within it. This kind of open piece, therefore, requires no title or interpretative "key." It is up to the viewer to make sense of it or to transfer onto it her own personal imagination; essentially, the viewer is the dominant creator of meaning. The origin-free work is not the expression of an artist's personal inner life or experience but rather forms a possible portal to the world (Ørskov 1987: 20–21).

In contrast to the international style of Bauhaus, the minimalist work of art is solidly anchored in time and place. But whereas the erasure of time and place, as part of the mission of the international style, is about what could be called a cultural and/or a historical time and place (a cleansing of all symbolic ornamentation and contemporary decoration), minimalism is concerned with being physically present in the here and now. The minimalist work of art exists in the present with the viewer. Experiencing it is, or should be, corporeal and momentary; it is this emphasis on presence that turns minimalism into a kind of universal mode of expression—or rather, it is an expression that can be read by everyone regardless of time and place.

The desire of minimalism to "force" the viewer to be present and experience the work without any interpretative filters (which are typically culturally contingent) anchors the work in a time and a place at the same time as it removes the time- and place-based knowledge parameters that often define our dealings with objects. This creates a sense of commonality or universality. We are all equipped with the same sense organs; we all have the same preference for balanced compositions combined with a desire to structure our physical surroundings. Therefore, in theory, we are all able to experience and consume—and derive pleasure from—the minimalist work of art.

Based on the international style's desire to remove itself from time and place (and thereby eliminate local differences) and minimalism's desire to create a sense of spatial presentness with the viewer, we can say that both movements, however differently, are characterized by a common quest for universal expression. Universalism is a mode of expression that affirms common aesthetic ground rules and that neutralizes local and contemporary parameters of understanding, and in so doing, might produce aesthetic pleasure for people regardless of time and place.

It is difficult to completely avoid creating a contemporary idiom—as happened with the Bauhaus designers and minimalist artists, equally. Their clean functionalism and pure cubes have in many ways become characteristic of the decades when they created their pieces, and this may be the curse of the artist (and designer). Everybody is a product of their time, and it is thus impossible to wholly escape the trends and movements of the era we are born into. Purifying the idiom and starting over by solely focusing on basic shapes and primary colors as one's only "building blocks"—like a child in kindergarten beginning by learning the letters one by one to finally being able to construct and read sentences independently—is not a real possibility. We all have "baggage" that influences both what we create and how we understand and interpret our surroundings.

### The universal effect of color

Concerning color, which of course is a vital part of an object's expression, it is possible to construct a set of universal guidelines that supersede symbolic, taught, and culturally contingent uses of color. The common experience of color is innate. Sensorially, human beings have a tendency, for instance, to regard darker colors as being less expansive than lighter ones; a white room seems bigger than a dark blue one; black appears slimming because our gaze decodes color in this way. Further, colors such as blue, turquoise, and cyan appear cool; whereas red, purple, and orange seem warm by contrast. A red chair will feel warmer to sit in or to stroke one's hands across compared to a blue counterpart. A room of cyan walls will feel colder than a room of purple walls, and so on.

Regarding color theory, a number of aesthetic principles challenging the symbolism of colors (red for love, white for innocence, green for hope, etc.) have preoccupied thinkers and artists such as Johann Wolfgang von Goethe (1749–1832) and

Johannes Itten (1888–1967). Decoding the symbolism of colors, which are largely culturally contingent, can only be done correctly by possessing an in-depth knowledge about the culture in question. This means that color symbolism is anything but universal.

Basic universal principles of aesthetics, in regard to color, can help us understand how different colors affect the human senses and how to create an expression that fosters the most harmonic composition. In his work of 1961, *The Art of Color: The Subjective Experience and Objective Rationale of Color*, Itten sets up guidelines for creating color harmonies or contrasts. For Itten, harmony means equilibrium or symmetrical composition. The eye is always searching for symmetry, according to Itten, and the color contrasts, when followed, can provide it with the means of achieving a sense of equilibrium and harmony, which is immediately pleasurable. For example, complementary contrast provides the eye with a calm aesthetic experience, as all three primary colors (red, blue, and yellow) are present, and because humans have an innate preference for the simultaneous presence of all three primary colors. This preference can be illustrated with the following example: staring for a minute at a green circle and then moving one's gaze to a white surface will result in the eye creating the illusion of a red circle (Itten called this illusion "simultaneous contrast"). Since green is a mix of yellow and blue, all three primary colors are here present.

The complementary and the simultaneous contrasts are two out of seven color contrasts that can help create different color harmonies. The rest are the contrasts of hue, light-dark, cold-warm, saturation, and extension (Itten 1997: 107). According to Itten, our senses can only apprehend and process objects and expressions by way of comparison, and the different contrasts can affect the senses in more or less powerful or diverse ways.

The harmonious composition of the color contrasts can nevertheless be challenged. Designers and artists can choose a challenging, dynamic, and vivid expression in order to change the color scale in relation to the contrast of extension and thereby make room for the most "voluminous" or expansive color (yellow, say) in a given composition, as this would break with the harmony, which is attained by maintaining the internal sense of proportion. Itten bases his concept of scale on Goethe's color theory.

Similar to Itten, Arnheim speaks of the weight of colors and about the importance of maintaining a sense of balance in a given composition by, for example, giving more room to certain colors over others and in this way meeting the universal human need for harmony and balance. As discussed, the human gaze seeks to structure its surroundings, and by keeping to the basic principles of aesthetics— regarding the weight of colors and the golden section—the practicing designer, artist, or architect can accommodate this need. In a sense, then, objects can be "charged" with the potential of meeting the innate human need and preference for harmony. And this is where we find the immediately accessible and instantaneous aesthetic pleasure—a kind of pleasure that appears when the viewer is presented with an idiom that is easy to apprehend and appreciate since the senses have been "programmed" to find pleasure there.

In line with Arnheim and Itten, Russian painter and art theorist Wassily Kandinsky has discussed colors and their physical effect on the human senses in his poetic and theoretical work, *Concerning the Spiritual in Art*. According to Kandinsky, colors have a physical, almost tactile, effect on the viewer that is based in the "spiritual vibration" of a given color (Kandinsky 2008: 59). This sensorial, synesthetic approach to color, which is characteristic of Kandinsky, is expressed in the following quotation:

> Many colours have been described as rough or sticky, others as smooth and uniform, so that one feels inclined to stroke them (e.g., dark ultramarine, chromic oxide green, and rose madder). Equally the distinction between warm and cold colours belongs to this connection. Some colours appear soft (rose madder), others hard (cobalt green, blue-green oxide).
>
> (Kandinsky 2008: 60–1)

The extent to which human beings are susceptible to the "vibrations" of colors and shapes, Kandinsky suggests, depends on their *spiritual sensitivity*. This means that the thing-in-itself contains a measure of or potential for beauty; however, it is only possible to truly impact the "spirit" of the viewer if she engages with and remains open to the aesthetic experience.

My treatment of Kandinsky brings us back to the quote from Dorthe Jørgensen that concluded the previous section, in which she points out that beauty "exists in the meeting point between a potentially beautiful object and a subject who, by virtue of her gaze, makes possible the experience of beauty as such" (Jørgensen 2012: 35). But what characterizes such a subject or human being—a human being who, to quote Kandinsky, is frequently met with "[t]he expression 'scented colours'" (Kandinsky 2008: 61)? And can spiritual sensitivity be taught?

Imputing to the artist (or designer) this kind of task is certainly idealistic: teaching humanity about spiritual sensitivity or about how to be open and receptive to aesthetic experiences and to recognize their value as such. Chapter 6 will return to this pedagogical issue concerning aesthetics.

Itten, Goethe, and partly Kandinsky take a phenomenological,[3] rather than a symbolic, approach to color. In other words, all three are concerned with how colors and their "weight" or combinations/contrasts affect the body (and the mind), how colors feel to be around, rather than with, what they symbolize or which associations they spark in the experiencing subject. The phenomenological approach ignores culturally specific symbolic values, which are changeable and which are thereby antithetical to the universal (unchanging, eternal) human experience of color. On the whole, phenomenology has a lot to contribute in terms of locating a universal experience of the world and its physical objects. If the corporeal understanding and apprehension of the world comes before the cognitive and the reflective, as French philosopher Maurice Merleau-Ponty (1908–1961) thinks, the cultural "baggage" and connotative framework of human beings are not necessarily an impediment to creating aesthetically durable expressions or objects with the ability to please or challenge recipients and thereby affecting them in powerful ways. In fact, cultural connotations are entirely insignificant in this regard.

German philosopher Gernot Böhme (b. 1937) writes about colors that they create atmosphere, that they communicate a certain mood in a room (Böhme 2010: 28). Colors can affect body and mind in specific ways, and since everybody is born with the same sense organs and the same body parts (more or less)—regardless of national heritage and time period—phenomenology can help us establish universal principles concerning the property and effect of different colors.

Similar to the phenomenological approach to color effects, the gestalt laws encompass universal principles for how humans (regardless of cultural background) experience and take in the forms of their surrounding world. For example, everyone will consider a composition based on the golden section harmonious. The use of certain materials can give the experience of an object a sense of "immediate correctness" and thereby a form of primary pleasure. If the material is tailored to the object shape, subtending its characteristics and functional properties, it can be easier to immediately "understand" the object and how it is used. As mentioned previously, Willy Ørskov operates with the term inertia, and it can be said that the less inert the material, the easier it is to apprehend, comprehend, and decode an object. The next chapter will bring further focus on the uncomplicated decoding of objects.

In line with the above, I have constructed what I call the "universal aesthetic principles." These principles should be followed when working with expressions that aim for an aesthetically beautiful, balanced, harmonious, and calm experience, which thus produce the Pleasure of the Familiar.

> It can be determined that the eye, which always tends toward simplification, has an innate ability to create coherence between forms and figures, but the interpretation of the emotional nuances between figures in the visual field is culturally contingent. This is an important point to keep in mind when decoding images from cultures and periods that are different from our own.
>
> (Gotfredsen 1998: 33; transl.)

As this passage from *The Idiom of the Image* [*Billedets formsprog*] shows, Danish art historian Lise Gotfredsen (1929–2009) claims that several elements factor into how we perceive and interpret images (and objects). On the one hand, there is the innate or universal tendency to seek out connections, and on the other, there is the cultural background of the viewer to consider. In the next section, I will examine more closely the cultural aspect of decoding objects, images, and buildings, as well as the sensorial "value" we add to our surroundings.

## The easily decodable

The phenomenological approach to taking in or apprehending the world and the objects of our surroundings is interesting in connection with the Pleasure of the Familiar and the universally most pleasing and beautiful expression. In many ways, this approach eliminates the taste-based, subjective, and culturally contingent elements of the individual's encounter with the world and objects, as the human being, according to phenomenology, is "reduced" to a sensing body.

However, it is difficult to deny that, as human beings, we impart the physical objects we encounter with additional meaning, or put differently, we attach a plethora of associations or connotations to them. A cup is not just a cup; a dress is not just a dress. A cup can symbolize artisanship, slowness, and *retro chic*, or it might ooze of mass production and anonymity. A dress can exude elegance and exclusivity, or it might emanate with a subdued sense of functionality and *substantialism*. These are all words and terms associated with cultural value, identity, and lifestyle.

When we consider physical objects or things as carriers of added value or different messages, we are effectively travelling in the realm of connotations, symbols, and myths, all of which belong to the science of signs: semiotics. Whereas phenomenology asks what it feels like to be present with objects and how they affect the mind and senses, semiotics asks how objects can be decoded or interpreted. Semiotics deals with associations and interpretations, as well as acquired knowledge and cultural signifiers, which have largely influenced our approach to the world. We are used to interpreting our surroundings because we have been taught that there is always something "more" behind what we see, hear, or read. It is a much more daunting task to consider a chair as being merely a physical thing that provides us with a purely corporeal experience rooted in either pleasure or discomfort, than it is to consider the same chair as being the *carrier* of status (or the lack thereof); stylistic or historical references; or trend-related symbolic value, for example. It is demanding to merely *observe* phenomena of the world without at the same time inscribing them with some kind of external value or meaning. As Willy Ørskov points out:

> To all extents and purposes, most people live their lives within a linguistic frame of comprehension regarding their surroundings. In keeping with this understanding, art-school professors consider it their finest mission to make their students see the world, train them in visual observation, seeking to replace their "knowledge" about the surroundings with a non-knowledge of the visual plane.
>
> (Ørskov 1999: 105, transl.)

In his 1964 essay, "Rhetoric of the Image," French philosopher and social theoretician Roland Barthes (1915–1980) discusses the "baggage" that determines how we decode and understand our immediate world and the objects within it. The connotations that appear to us when "encountering" a physical object are based in our cultural context, our beliefs, and our values, as well as our lifestyle and social circle—our connotative frame of reference. Understanding people's frame of reference is crucial for designers wanting to reach a certain demographic, or a certain segment, with their products.

In this context, it is useful to draw further on Barthes' essay, specifically concerning his terminology of denotation and connotation. Denotation refers to the basic meaning of any sign (an object or an image, for instance), but as Barthes points out, it only makes sense in theory to talk about denotation. In reality, there are no denotations without connotations. Denotation is the sign in itself—without any added value or meaning. Experiencing the world as a series of denotative signs—without attempting to interpret them or attach additional meaning to them aside from the fact that they consist of shapes, colors, and different materials—is impossible, according to Barthes (and semioticians more broadly). As human beings, we cannot help but attach meaning and value to signs as they come to us. This behavior has to do with how, in a semiotic optic, we structure and order our immediate surroundings. We filter and categorize everything we "meet" along our way through life by drawing on previous experiences as well as personal and social beliefs; this is the only way the world makes sense to us.

However, the denotative approach, if we can call it that, is similar to phenomenology in many ways. Basically put, to experience the world and its objects phenomenologically means experiencing phenomena for what they are without reference to any religious, cultural, habitual, or temporal framings, categories, or connotations. Connotations are very similar to associations, but whereas associations belong to the intimate sphere of individuals, connotations has to do with our cultural "baggage" and habitus, the frame of reference we share with other like-minded individuals.

Connotations refer to the meaning human beings apply to the world and objects. Connotations are the end result of the subject's interpretation of phenomena. As mentioned, connotations are linked to associations, but the former are rooted in the subject's background and lifestyle as well as the spirit of the time. For instance, present a contemporary urban dweller in the western world with a loaf of overnight-proof yeast bread on a rustic, wooden cutting board accompanied by a jar of homemade rosehip pesto, and this particular still life will most likely result in connotations such as "laidback luxury" and "quality" as opposed to "simple peasant food."

As the sender of a message (or as the designer of an object), it is possible to control or anchor the connotations evoked, to some extent. What Barthes calls "linguistic messages" can be used to attach certain words to the object or product (material or immaterial) before it is distributed. This could take the form of a product title or a neck label containing information about the concept or design process behind the collection; furthermore, online storytelling could be used as an anchoring tool to supplement the physical experience of the product.

Another way in which to anchor connotations in a product is to juxtapose it with visual elements. Returning to the image of the loaf of bread resting on the rustic cutting board next to the jar of rosehip pesto evoked above, the visual part elements of this still life (the bread, the board, and the pesto) complement and enhance the meaning of each other. This should be understood in the way that the gaze falls on the loaf of bread and the immediate connotation is affirmed (the slow pace and care in making the bread) as the gaze travels to and involves the cutting board and the jar of homemade pesto. The part elements captivate and hold the viewer's thoughts, so to speak, preventing them from wandering off to some unintended place.

In other words, what Barthes calls "anchorage" (Barthes 1977: 38) has to do with making sure that the intended understanding or decoding takes place. Working strategically with anchoring the viewer's connotations (by carefully considering both the linguistic messages and the visual part elements of one's product, which may be presented online or in a physical collection) is therefore crucial to the design process or the planning of a product launch.

In the world of semiotics, everything is a sign. All objects contain meaning and messages, and a message only affects the recipient if it makes sense to her. It is therefore important that the sender (and thus the designer) understands the recipient's, or potential customer's, connotative frame—meaning, what a given culture accepts as the constituent truth. Hence, in relation to strategic design considerations, it is imperative that designers thoroughly research their target audience in order to tailor specifically the desire evoked by their products. What do potential clients or customers believe in? What are they influenced by? What do they find interesting, beautiful, exquisite, trendy? And what structures their consumption? The connotative frame encompasses the answers to all such questions, and the frame is highly contingent on social meaning exchanges and cultural convictions.

In his book *Organizational Culture and Leadership*, American social psychologist and cultural researcher Edgar Schein (b. 1928) employs the term "assumptions" (Schein 2004). Assumptions are the "invisible," unconscious opinions and views on life that order or are at the root of a group's behavior and their shared culture. Assumptions are anchored so firmly that it can be difficult for group members to understand behaviors and lifestyles different from their own. In this way, assumptions are "commonplaces" or "truths" that are difficult to challenge, and that exceedingly determine the consumption patterns and preferences of the group members.

Schein's assumptions are similar to Barthes' understanding of connotations, and further, Schein's theoretical framework reminds of another Barthesian concept: cultural myths. By myths, Barthes means the stories or "truths" that people of the same culture believe in and live by, and which determine the way they decode their surroundings. An example could be the myth of "the one and only." The basic assumption or myth about "the one and only" dominates most Hollywood blockbusters and western romantic stories. But this myth, that we all have just one perfect romantic partner, is completely foreign to cultures where arranged marriages are the norm. In such cultures, the basic assumption is that marriages function well because the spouses are determined, and because they match the plans, norms, and wishes of the families involved.

Religions are brimming with myths. Consider, for example, the Christian myth of Adam and Eve and their fall from grace. This kind of mythological story is a "truth," which means that it is truthful to those who believe in it. Put differently, it doesn't make sense to question its veracity or, referencing Darwinian theory, to challenge the assumption that the Garden of Eden was a real place whence the origin of humanity and suffering can be traced. Myths or assumptions are in a sense not up for discussion. They negate all basic theories of argumentation. Either you believe, or you don't.

Common to all myths and assumptions is that they are so deeply essential to being human that they are considered "capital-t" Truths, and for this reason, they can be difficult to argue against or explain away to those who believe in them ("it's just the way things are"). When researching a specific target group for one's design or product, the primary assumption of the group will reveal itself as they run out of rational explanations to argue for preferences and dislikes: "Handmade things are nicer because . . . well, just because they are"; or, "one's line of work should be a reflection of one's passion, that's a given." Counter arguments—such as, "Handmade things are uneven and full of little mistakes, making them less good than mass-produced things"; or, "It makes more sense to work with something that's less all-consuming, making it easier to separate work from leisure and family life"—will typically have little effect on someone who is convinced and feel that their assumptions are a fundamental part of their identity.

Gaining knowledge about the assumptions of one's target group is of immeasurable value. If, for instance, one's target group considers handmade objects inherently beautiful, as a designer, one's job is to analyze why this is so. A good starting point in this regard is to examine contemporary myths or the assumptions that flourish and characterize the "mythological sphere" or community of meaning the recipient belongs to. In the particular example of handmade objects, it could be that artisanship signals humanity and integrity, time and resourcefulness, and that these values are rare commodities in our day and age. For this reason, they are likely to be considered exclusive and thereby beautiful. Having reached this conclusion, explaining the values and needs of the recipient, the next step is to figure out how one's concrete design or product can somehow accommodate these. Perhaps slowness and the passing of time can be emphasized as part of the design's expression by incorporating handwriting or another form of "fingerprint," or perhaps a storytelling that focuses on the hands and labor behind the concept can make the design stand out in this way.

If, as a designer, the aim is to create a product that triggers the Pleasure of the Familiar in the recipient, one must, in this semiotic lens, incorporate a symbolic value into the product that can be easily decoded. The Pleasure of the Familiar is not caused by complex linguistic messages or Dadaist juxtaposing of visual elements, but rather by an immediately decodable idiom and by familiar references that evoke the recipient's connotative frame, myths, or assumptions. In other words, the recipient's assumptions must be either affirmed or supported by one's design.

Familiar, or "homely," references can be used to create a type of additional value that can be immediately and easily decoded and appreciated. The recipient must be accommodated and satisfied within her comfort zone—individual assumptions and presumptions must be affirmed, but this is only possible if the sender (the designer) has done her or his job properly by researching and analyzing the recipient's Truths about the world. The comfort zone, put differently, must be "padded" and all cracks filled, so that nothing might feel uncomfortable or foreign; the recipient's community of values must simultaneously and immediately be able to understand the object's significations. The aesthetic experience has to be familiar, comfortable, affirming, and straightforward.

It is thus crucial that when working with the Pleasure of the Familiar, one understands how to support, rather than to challenge, the recipient's assumptions about the world. The mode of expression and choice of materials must be exactly what they "usually" are, meaning that a clothing design, for instance, should not contain any obscure lines, seams, or non-functional elements; a piece of furniture should not challenge our usual (in the west) way of sitting; titles should not distract from the use or function of the design, which would prolong the decoding process; materials should not confuse the senses by producing a cool instead of a warm impression in a context where the latter would be expected; and so on. The symbolic value of the design should be clear and easy to decode; the aesthetic expression of the object should refer to the myths and values that are fundamental to the target group. Achieving this can help create the kind of pleasure that is characterized by immediately feeling at ease, comfortable, understood, and affirmed in one's beliefs.

## Homeliness

Homeliness, to me, is going for a walk on a cool and windy autumn day, keeping in my jacket pocket a chestnut that I form my grip around, feeling how it slowly absorbs the heat of my palm. There are many parts to this overall feeling of homeliness, but common to them all is that they are connected to my body: the smoothness of the chestnut against my palm, the warmth of the pocket in my wool coat, the rhythm of my gait, and the sharp cold of the autumn wind against my face. These all combine to form a tactile experience, or they could be described as a tactile sense of home. They create a tableau of harmony and anchorage. Using a somewhat awkward term, "the familiarity factor" is highly present. In continuation of the section on the universal aesthetic principles, there is a great sense of pleasure in experiencing how the physical world "behaves" exactly as we would expect. A comfortable and homely sense of pleasure is characterized by not being demanding or challenging, but rather affirming and acknowledging. In other words, we are here dealing with a kind of pleasure that supports the ideas and expectations of our surroundings and the objects that make us feel at home in the world. The experience of homeliness gives us the impression that the world is a familiar, predictable, and comfortable place to be. Homeliness is a saline solution injected into our comfort zone.

The above description of the chestnut-in-the-pocket experience is an aesthetic experience in the sense that it is pleasurable and tactile; it makes everything assume a greater sense of purpose and being—or simply, belonging. In his collection of poetry *The Worst and the Best* [*Det værste og det bedste*], Søren-Ulrik Thomsen writes about a similar experience: " . . . when a beautiful grey suit makes my devastating sense of unease possible to bear, and God's silence shuts out the roar of the bad blood, that's the best" (Thomsen 2002; transl.). This sense of coherence is often tied to memories as well as corporeal or sensuous experience. Or, drawing on Thomsen, it could be connected to the experience of a certain artifact, as for example a beautiful grey suit or a snug wool coat with large, spacious, warm pockets that makes it possible and bearable to enter the world on days when everything feels unmanageable. What we wear or surround ourselves with adds to our sense of identity, supports and grounds us.

The sensuousness of the experience (the smooth surface of the chestnut against the palm; the hands in coat pockets; the wind against the face; the feeling of feet in boots; and the rhythmic, meditative walking), which in this case is connected to tactility, but which could just as well concern taste, hearing, or smell, cancels out the experience of time passing. The past and the present merge as a moment takes hold of us, becoming eternal and timeless. It is like a refrain recurring throughout life that binds together the seemingly insignificant events and disjointed "stanzas" that a life consists of. In such moments, we become one with the world; the boundary between self and world is momentarily suspended and a clarity of vision follows; or rather, a clarity of "feeling," a feeling of purpose, coherence, and cohesion. A religious person would probably, without hesitation, describe such an experience as a revelation.

The elements of the sensuous aesthetic experience and feeling can best be described by a quote from volume 1 of *In Search of Lost Time* by Marcel Proust (1871–1922). The pleasure and joy of life, as well as the sense of coherence, that Proust describes here is characteristic of the "refrain experience" I earlier called a chestnut-in-the-pocket experience:

> But at the very instant when the mouthful of tea mixed with cake crumbs touched my palate, I quivered, attentive to the extraordinary thing that was happening inside me. A delicious pleasure had invaded me, isolated me, without my having any notion as to its cause. It had immediately rendered the vicissitudes of life unimportant to me, its disasters innocuous, its brevity illusory, acting in the same way that love acts, by filling me with a precious essence: or rather this essence was not merely inside me, it was me. I had ceased to feel mediocre, contingent, mortal. Where could it have come to me from—this powerful joy? I sensed that it was connected to the taste of the tea and the cake, but that it went infinitely far beyond it, could not be of the same nature.
>
> (Proust 2004: 98)

As mentioned earlier, there is something "universal" about the corporeal experience of the world that, to an extent, eliminates cultural and social differences. We are all equipped with the same sense organs, and we thus experience—corporeally— our surroundings and physical objects in largely the same way. This corporeal or sensuous aesthetic focus contains a universal human commonality, which is significant in relation to the concept of sustainability, as it disregards subjectivity in favor of the potential for creating guidelines of forming pleasurable objects that transcend time and space.

The memories I have of my childhood home, the feeling and sense of being there, are largely physical in nature. In particular, my memories are connected to tactility, my sense of touch: I remember the *feeling* of walking through the house.

In a more concrete sense, I can remember the changing texture of the surfaces under my feet: sisal, wood, and tile—my memories are clear enough that I can recall the scratchiness of the sisal rug, the smoothness of the hardwood floor, and the coolness of the quarry tiles against my bare feet (based on this, I've always preferred walking in bare feet). Recalling these sense impressions brings up different moods that I would not otherwise have access to. These kinds of memories could be called sensuous experiences:

> The sense of home is not just a question of articulating the stories and memories that are attached to a particular place. Memories and subjective experience are articulated through sensuous events—smells, creaky stair boards—and how different places, in particular the home, impress upon us.
>
> (Bille and Flohr Sørensen 2012: 105; transl.)

Physical memories are wordless; they are vehicles for time travel, triggered in the blink of an eye by similar physical experiences. Such memories can lead to a plethora of insights, associated thoughts and feelings.

## The experience of minimal inertia

As a designer, "playing up" the physical, sensuous power of memory (while also focusing on the Pleasure of the Familiar) can be done by creating objects where the relation between material and form is one of minimal inertia. The concept of inertia, in this regard, comes from Willy Ørskov (1987). Concerning materials, inertia refers to how resistant they are to being molded a certain way that will make them "fit" the form of the object they are to become. If a material is inert, it is the opposite of compliant, so to speak, meaning that it will have to be processed before it can be made to fit the mold of the object. Conversely, a material of minimal inertia will easily assume the form of the object. It will be easy to shape into the desired form. It complies with the object.

Observing or touching an object whose material is of minimal inertia, one will immediately understand (using one's senses) how the material came to take shape as such. There is nothing to wonder about and the material thereby does not form part of the apprehension process. Ørskov is a phenomenologist, and it is important for him to point out that when we "encounter" objects using our gaze, body, and senses, we focus on *detecting* as opposed to *decoding* them (see the section on semiotics above):

> Detecting an object is markedly different from judging or interpreting it. Whereas judging, or categorizing, the object concerns its gradation and numbering in relation to other objects in a pre-established value system, appre- hending presupposes a kind of non-engagement and a disregard of value systems. [ . . . ] And whereas interpretation makes the object into a point of attachment for meditations, dreams, associations, references (a psychoanalysis

of the object), the act of detection does not supply anything external to what is already materially present in the object.

(Ørskov 1987: 77; transl.)

Detecting an object, according to Ørskov, means being present with it, sharing a space and experiencing it, without at the same time ascribing any added immaterial meaning to it, burdening it with culturally contingent values and subjective con-notations. A chair is not a status symbol or a sign connoting "trendy substantialism" or "anonymous mass production." Rather, it is simply a chair consisting of a base, structure, and support, as well as a more or less inert material, which may appear either anonymous or domineering in terms of the idiom. Ørskov refers to the fact that a material appears in its purest form as "material romanticism" (Ørskov 1987: 90). If the wood material is painted or upholstered, rendering the pores invisible and the surface inaccessible to touch, the material has been "anonymized." Material romanticism, instead, would seek to emphasize the pores and the intricate surface of the wood as an essential element of the object's expression. Of course, painted or upholstered wood can still be read as being wood, but it is nevertheless easier for the senses to immediately detect the material if, according to the romantic material-ist approach, the characteristics of the material are allowed to appear on their own.

Designers wanting to provide their recipients with the type of aesthetic experi-ence that evokes the Pleasure of the Familiar, and thus a sense of homeliness, will have to decide on the extent to which material and form go together (as part of the design process). How "easy" is it for the material to accommodate the form demanded by the object's expression? To allow for the subject to experience a sense of minimal inertia when detecting the object, as part of the easily accessible kind of enjoyment associated with the Pleasure of the Familiar, the object's form/ expression must figure in an immediately visual and tactile symbiosis. By this I mean that the immediate experience of the object must not be ambiguous or obscure in any way, regarding material and form; the choice of material should be obvious in relation to the form of the object. Having said that, the chosen material can still be of the anonymous sort as long as it is the best suited for the kind of dress, chair, boot, or plate one is designing. For example, in the case of a jacket-object, the material should be soft (wool, suede, or certain textiles) and match the jacket's form and main function: to cling around the body. One would therefore not choose to construct a jacket from plastic, wood flakes, or wickerwork. In the case of a table, it must be made from a solid material (wood, metal, concrete) that will be able to withstand pressure, and not a foam material or knitted elements that would soften the hard surface.

Additionally, the material, to exhibit the principle of minimal inertia, must be easily made to fit the form of the object. It must be *compliant*. This means that if the form of the object is curved, the material must "adopt" this form in an uncompli-cated manner—bending to the will of the object, so to speak. At the same time, it is important that the material can be read as being pliable. For example, if a given chair is meant to consist of sharp edges, non-pliable materials such as wood or metal should be used. But if the chair, on the contrary, is meant to evoke organic forms and arched lines, plastic or soft chipboard should be used, as these materials give the impression that they have been made to suit the form of the object perfectly; accordingly, the recipient will experience the object as one of minimal inertia.

The experience of minimal inertia entails a sense of the material as appearing with all its natural characteristics and potential uses. It will not have been manipulated, processed, or made to fit a form—at least not to a significant degree. It is left to "act" as best it can. A material of minimal inertia can still be *anonymized*, but in terms of form, the experience of minimal inertia is characterized by a sense that the material has been allowed to assume whatever form might best suit it. From the perspective of the experiencing subject, it might almost seem as though the material has guided the designer or artist, and that as part of the creative process, it has "told" her which form suits it best or which form it can "assume" in the least resistant way.

To enrich and acknowledge the expectation of its recipient by drawing on the Pleasure of the Familiar, an object must be easy to detect. The experiencing subject will need to have affirmed its bodily or sensual assumptions about how the object "acts" or feels. How does it feel to run one's hands across its surface? How does it feel to lift? Is it heavy or light, massive or pliant? The experience should not be colored by tactile and visual surprises or by an aesthetic that breaks the comfort zone of the recipient.

The human power of memory and the ability to create connections between events and objects are related to the experience of feeling at home. Our cognitive abilities can connect apparently insignificant (or equally common) and random events to create a beautiful and motley patchwork of asymmetrical import. Meaningfulness—even as a momentary and fleeting experience—is crucial to feeling at home in the world. This is exactly why it is characterized by the sort of aesthetic experience that is linked to the Pleasure of the Familiar. Things carry memories, and they can help create a sense of home; we surround ourselves with beautiful and comfortable things in order to build mood and to emphasize who we are and what we stand for. Moreover, we tend to care especially well for those things that have special value to us (what we could call sentimental value) because such things evoke mental states that we are fond of inhabiting.

Things that make us feel at home, and which carry memories, are magical; they have a certain *aura*. In later chapters, I will come back to the concepts of aura and magic, which are often related to feelings, moods, and memories. However, magical objects are also highly significant in relation to the bodily and sensuous feeling of home. Like a chestnut in one's pocket, magical objects can create a sense of comfort and safety, which characterizes feelings of belonging and presentness.

Not all objects are supposed to challenge our senses. Only insofar as it makes sense should design objects challenge the sensorial apparatus of their recipients. Unnecessarily challenging designs will only irritate recipients. For example, most kitchen utensils and aids, which are considered functional objects in the main, should trigger in users the Pleasure of the Familiar. The form of these kinds of objects should accommodate expectations of how they are supposed to function, and the design material ought therefore to facilitate the form and appear as being minimally *inert*. The purpose of kitchen objects should thus be easy to detect, preferably immediately, so that they can be used without problem. If the designer of a mixing bowl, for instance, wishes to challenge the user, this should be done on a minimal scale—perhaps by "twisting" the form ever so slightly or by challenging the material with a view, preferably, to improve upon the user's ease of mixing ingredients. The object of the mixing bowl should remain as useful as initially conceived (cf. Plato's conception of the most beautiful spoon as being the spoon that is the best at being a spoon), unless the intention is to create a statement or sculpture that functions to give identity or status in the kitchen. If the latter is the case, we are then dealing with the kind of pleasure that in the next chapter will be referred to as the Pleasure of the Unfamiliar.

## Conclusion

The type of aesthetic sustainability associated with the Pleasure of the Familiar can be described as that of which one never tires, like the magical things that follow us through life, or like the beautiful objects we have always had and continue to care for because they suit everything else in our lives. These kinds of objects complement our sense of self (or who we want to be).

The Pleasure of the Familiar is related to the beautiful aesthetic experience as it contributes an immediate feeling of comfort to the experiencing individual, who is then affirmed in the belief that everything in the world is exactly as she had expected. In the Pleasure of the Familiar, we encounter ourselves. This might sound strange (because how is it possible to encounter one's own self?), but the idea is that we are "confronted" with ourselves—what we stand for, are, and contain as people—and not least of all the expectations and assumptions that we hold about the world: the basic assumptions of Schein. The aesthetic experience affirms one's individual sense of self, the feeling of belonging in a certain place and with a certain group of people. On the one hand, this can be understood in a purely phenomeno-logical way, as in how the aesthetic experience affirms the corporeal and sensuous expectations we garner in relation to our surroundings and the objects we encounter. On the other hand, it can be understood purely symbolically, as in how when encountering objects or concepts that trigger the aesthetic experience we are confirmed in our culturally and socially anchored assumptions about the world.

## Notes

1  The use of the term inertia refers to Willy Ørskov's use of the term in *Detecting Objects* from 1966. Ørskov's theory will be further elaborated later in the book.
2  The Bauhaus School was a German academy for design, art and architecture, established in Weimar in 1919. In 1925 it moved to Dessau, where existing until 1932. Hereafter, the school moved to Berlin, where the Nazis closed it down the following year.
3  Phenomenology generally seeks to detect the physical, sensuous qualities of events and objects, rather than what they symbolize.

# 2   The Pleasure of the Unfamiliar

In many ways, the Pleasure of the Unfamiliar represents the antithesis of the beautiful aesthetic experience; in other ways, it can be seen as its prerequisite. It makes no sense to discuss either kind of aesthetic pleasure in isolation. Challenging one's boundaries or being *pushed* towards new experiences can be highly satisfying *and* deeply uncomfortable at the same time. Similarly, the prospect of taking a new turn in life can be both seductive and terrifying. In any case, overcoming obstacles by superseding one's own expectations is empowering.

We take many things for granted in sensing the world: for example, the fact that tables have cool, smooth surfaces; that pants have two legs; that rings are round; that chairs have four legs; that hats are worn on the head; that dresses must have a flattering silhouette; that cups hold hot liquid and feel even and smooth against the lips. When confronted with an object that challenges conventional expectations, we are forced (if just for a moment) to stop and think about what is going on. The duration of the moment is linked to the object's degree of complexity. Whether the moment consists of minutes or seconds, the experience is set to challenge and push us in different ways. In this way, the basic assumptions that govern our interaction with the objects that surround us are stretched, distorted, and expanded—maybe only to return to the same familiar form momentarily. Or they may be altered forever.

The kind of aesthetic experience that is connected to complexity and unusual combinations—which the following sections will cover—grabs us, shakes us, and intentionally challenges our expectations. If the challenge isn't intentional—if we are confused by the object; if we can't figure out how to use it; or if the material is inappropriate given the context of the object's possible uses or its idiom; and if, at the same time, the object isn't stimulating or aesthetically nourishing—we are not dealing with an aesthetic experience, but rather a poor design experience.

Some objects can't immediately be understood or decoded; instead, they challenge our immediate need to conceptualize, generalize, and organize—and the encounter shocks our imagination. Encountering complex objects or objects that consist of unexpected combinations—or that to a lesser or greater extent play havoc with our expectations of the world—we are forced to stay with the present moment to capture and understand what greets us. Our consciousness is pushed to "expand."

To a certain extent, challenging experiences are uncomfortable—but this sense of discomfort also contains a special kind of pleasure. Human beings have a fundamental need to be challenged and pushed. This existing need might be greater, lesser, or equal to the need for safety, order, harmony, home, and the "familiar." Based on the fact that the need to have one's assumptions challenged on occasion is fundamental to the human experience, the Pleasure of the Unfamiliar can be said to be "durable" or sustainable. The challenging (design) object can therefore be called aesthetically sustainable.

Unchallenged, the human spirit stagnates. Human beings can't improve or evolve without being challenged. Challenges force us to form, transform, and re-form our assumptions about the world, thereby expanding our consciousness and horizon. The challenging aesthetic experience, which can lead to the Pleasure of the Unfamiliar, has a lot in common with the sublime aesthetic experience. The following section will therefore revolve around the concept of the sublime.

According to the German philosopher Friedrich Schiller (1759–1805), experiencing only beauty and harmony—without the contrasting experience of chaotic phenomena characteristic of the sublime—fixes the human to concrete, perceptible reality (Schiller 2010: 128–29). Clinging to the familiar, the safe, and the immediately understandable, without seeking out challenges, leads to stagnation—and in turn, un-freedom. Only the sublime, which Schiller connects to chaotic or unbounded experiences, can stir human beings from their slumber, reminding us of our full potential. It is thus important to occasionally be confronted with the chaotic elements of reality. A single sublime shock can tear apart what Schiller terms the "veil of illusions", creating for the individual a momentary experience of *true beauty*—meaning the beauty of the idea, not that of the body or that of sensuous, physical reality (Pugh 1996: 117). *True beauty* merges with the sublime in the sublime aesthetic experience. Experiencing, for example, immeasurable heights or boundlessness can lead to a confrontation with something that goes beyond our immediate sense perceptions, which is why such experiences can seem liberating. The particular value of the sublime is founded on its ability to challenge and provoke the human experience, shaking us out of the passivity related to that which is familiar and safe.

It is possible for everyone to experience beauty and also sublime moments, but human beings, according to Schiller, can be *educated aesthetically* to become more receptive to such experiences. And the artist can spearhead this form of aesthetic education. In this way, Schiller attaches to the artist an idealism that can be transferred onto the designer. If the designer can contribute to the aesthetic education of human subjects, thereby making them susceptible to aesthetic nourishment, the designer assumes a responsibility that is closely connected to the aforementioned notion that the primary function of design objects is not necessarily exclusively practical or needs-based.[1] In other words, the primary function of design objects can certainly be aesthetical.

The sublime aesthetic experience leaves traces in the experiencing subject's mind long after it is over—and this in spite of the momentary nature of the experience itself. The experience, so to speak, tugs at the subject's sense of the world, to a lesser or greater extent. In great and small measure, equally, the sublime aesthetic experience touches the experiencing subject. The shocks to the soul that accompany sublime experiences can be relatively minor in scale, and might simply result in an experience of being "prodded" ever so slightly; sublime shocks might also be of a more fundamental character, contributing to lasting changes in the subject's mind.

This increase or decrease in the intensity of the sublime experience can be illustrated by Figure 1.

The "mild" sublime aesthetic experience (being prodded)

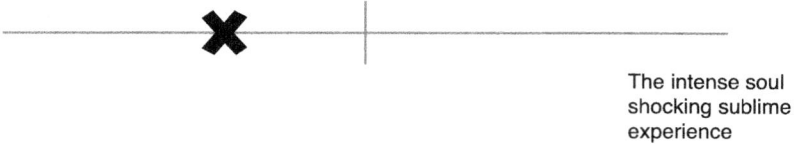

The intense soul shocking sublime experience

*Figure 1* Conjugating the sublime aesthetic experience

Designers ought to locate the point on the scale that makes the most sense in relation to the product category in question and the desired or most likely recipient of the design. If designing a kitchen tool, for example, it probably will not make much sense to strive for an intense, soul-shocking sublime experience. On the contrary, if the goal is to "disturb" slightly the recipient's basic assumptions about and usual approach to kitchen tools, it could make sense to create objects that can be immediately and easily decoded and used, but which nevertheless are shaped using unconventional materials or appear in a form that differs from a "normal" wash basin, for instance, and which will accordingly prod the recipient somewhat, forcing her to pause for a moment during her daily routine.

Designers must be familiar with the universal or common aesthetic rules (symmetries, color harmonies, the golden section, material experiences of minimal inertia, etc.) and be practiced in employing them before it becomes possible to break them. As stated in Chapter 1 concerning the Pleasure of the Familiar, it is possible to determine a set of guidelines for how to create a design object that meets the basic human need for order, structure, and comfort. Such guidelines form a crucial fundament for embarking on a project of breaking with immediately comfortable aesthetic expressions and thereby challenging user expectations. The following discussion of the sublime and the Pleasure of the Unfamiliar will demonstrate how to intentionally challenge and break the universal or common aesthetic rules to thereby accommodate the antithetical, but nevertheless human, fundamental need for disorder and chaos.

## The sublime

As already mentioned, the difference between the beautiful and the sublime concerns the difference between order and chaos; between symmetry and asymmetry; between predictability and unpredictability; between demarcation and boundlessness; between form and formlessness; between proportion and irregularity; and, finally, between the kind of aesthetic experience that nurtures one's comfort zone and that which challenges or breaks it. The beautiful and the sublime are at once diametrically opposed and mutually dependent on each other. In a sense, they embody the yin and yang of aesthetics. They are fundamentally different, but existentially dependent. For instance, it is nonsensical to speak about symmetry without at the same time understanding the concept of asymmetry, just as harmony cannot be grasped completely without its opposite, disharmony.

In 18th century aesthetic and philosophical treatises, the idea of the sublime comes to the fore in a big way. The sublime is generally considered as a counterpoint to classic beauty, as discussed in the section on the beautiful in the previous chapter, and thereby as something formless, chaotic, horrific, and alien. The sublime is the antithesis to the proportioned, symmetrical, and elegant. The first to name the difference between beautiful and sublime aesthetic experiences was British critic Joseph Addison (1672–1719). However, Aristotle (384–322 BC), in *Poetics* from around 335 BC, had already called attention to the multifaceted nature of aesthetic experiences. Using the concept of catharsis,[2] he discusses the nature of aesthetics, touching upon the curious phenomenon of how human beings are drawn to moments of emotional release—such as sobbing during sentimental plays (or films)—even finding comfort in doing so. Although of a different affective nature, experiencing shock or disgust in the face of terrifying art pieces is also considered cathartic in the Aristotelian optic. Clearly, then, the terrible or terrifying consist of a kind of pleasure that is markedly different from "pure" enjoyment.

Briefly put, catharsis names the effect that a well-composed tragedy has on its audience. It is a form of purification of soul or mind, connected as it is to pity and fear (Aristotle 1996: 32), which entails a transformation from intense and painful passion to a balanced, calm, or elevated frame of mind. For catharsis to take effect, Aristotle says, there must exist between the audience and the tragic performance itself a certain distance, an appropriate distance that is neither too small nor too great. In a dramatic situation of close proximity between audience and stage, the audience will be drawn into the action to such a degree that they forget that they are in fact safe from the action on stage; this will cause the audience to lose themselves in the play through an all-encompassing sense of fear and a corresponding feeling of pity for the hero of the play. However, if the distance between audience and stage is too great, the heart will not be moved—the experience will seem insignificant and ineffectual. The proper aesthetic distance, on the contrary, involves the human audience (the spectator of tragedy in Aristotle's world) emotionally, but only in an observational capacity (Scheff 1979: 59–61). The *soul-shocking pleasure*, which reminds of the sublime aesthetic experience, can only come about if the experiencing subject's distance to the overwhelming action is maintained.

Through the "good" tragedy, a transformation from pain to an elevated, balanced frame of mind occurs in the spectator. The tragic spectator can experience all kinds of adversities: unhappy love, birth, death, hate, lost love, grief, etc., without giving in fully to the power of emotion. From a safe position, comfortably seated in a theatre row, the on-stage troubles can be followed, experienced, challenged, tested, and felt, perhaps (preferably) aided by tears. And when the cathartic end of the tragedy has been played through, the spectator is free to stand up and venture out into the world, rejuvenated by the fictitious human experiences just witnessed, with a purified soul and mind.

Here we find an essential point that is closely related to the sublime aesthetic experience and the Pleasure of the Unfamiliar: the experiencing subject must be able to find her way back *home* again. The point of the experience isn't to lose oneself entirely, or to let go of everything and give in to the seductive pull of ecstasy. The sublime aesthetic experience only affords a *momentary* loss of self; it is a brief loss, which will be recovered in due time.

### Order–chaos–order

The sublime aesthetic experience contains a process, or a structure, which is not unlike the classic *Bildungsroman* progression of "home-away-home"; thus, the sublime experience is structured as follows: "order–chaos–(new and improved) experience of order". In each case, there must be some kind of *payoff* or reward in the end. Having been shaken by the forces of chaos, one needs to be guided back to safety ("home") again after an appropriate duration of exposure.

In *A Philosophical Enquiry into the Origin of our Ideas of the Sublime and Beautiful* from 1757, Edmund Burke (1729–1797) connects the sublime to vast, redoubtable (nature) experiences. Thus inspired, one could depict a scenic, sensuous image to illustrate the sublime aesthetic experience, which, customary at the time, was bound to nature. The scene unfolds like this: A wanderer is climbing a steep mountainside. At long last, he reaches the top; he is met by a glorious, breathtaking view, which makes him feel infinitely small. The sheer size and force of nature is overwhelming. Suddenly, dark, threatening clouds appear on the horizon. The wanderer immediately seeks refuge in a rock-side cave, from which he can safely observe the scene unfolding outside. The storm hits. Rain, hail, and lightning erupt from the sky. The wanderer senses immediate danger. The situation is overwhelming—so overwhelming, in fact, that it is paralyzing. But, at once, the threat seems to dissipate. He realizes that he is not actually in any physical danger; the cave shelters him, and the storm will eventually pass. Instead of alarming, the power of nature now seems fascinating. Relief and a sense of calm wash over the wanderer. His senses, which only moments ago were in a state of paralysis, are beginning to intensify. Different smells, sounds, and sights all contribute to a new experience of self, intensified by the shift in mood. Using the power of reason, the wanderer has triumphed in the face of danger.

Structurally, the above depicted narrative of the sublime moves from order to chaos and then, lastly, to a new and improved sense of order:

1   Order: The mountain climb is described as the starting point of an experience of being in control; the wanderer has just conquered a steep mountainside and is confronted with a glorious view of the landscape below.

2    Chaos: Soon after, however, chaos hits. The tempestuous storm leads to a brutal break with the wanderer's comfort zone—followed by a seeming loss of control and self.
3    (New and improved) Order: Having sought refuge in a cave, the wanderer experiences a newfound feeling of safety, and this experience engenders a blissful sense of being able to rationalize what had just seemed so dangerous and terrible. The senses are intensified, expanded, stretched—and a feeling of being fully alive and present sets in. The *momentary* loss of self, then, had the function of catharsis, purifying and evolving the self.

This three-step process accurately summarizes the sublime experience.

In relation to the aesthetically sustainable design experience, the Burke-inspired take on the sublime can be "translated" as follows:

Imagine standing in front of an armchair. It looks cozy, soft. It appears to be upholstered, an assumption that its warm plum color immediately supports. It looks as if it has been covered in a material similar to wool. However, stepping closer to touch the material, your immediate haptic expectation is thwarted. As your fingers draw across the chair's surface, you feel confused and disoriented—the chair is cold and rough. Leaning against it, you realize as well that it is much heavier and more substantial than expected. In fact, it does not budge. You begin exploring it in more detail with your hands. The material is not immediately familiar. You spend more time with the chair. Sitting down, you are met with a hard, cold, and rough sensation. After some time, it occurs to you that the chair is made from dyed to match molded concrete. You leave this experience with the knowledge that chairs can *actually* be made from concrete, imparting a corporeal, sensuous experience of heft, roughness, and coldness.

The above encapsulates precisely the Pleasure of the Unfamiliar.

### Sublime terror

In the 18th century, philosophical, aesthetic thoughts flourished about the nature of beauty and the aesthetic experience. Burke suggested that the classical ideal of beauty—as pursued by neoclassical artists, who (referencing the Hellenistic world-view) adhered to symmetry, harmony, and order—no longer included all facets of the beautiful. Burke used the concept of the sublime to connote everything that evokes aesthetic pleasure, but which falls outside the sphere of classical beauty.

To distinguish between the beautiful and the sublime, Burke operates with two basic affective groups: pain and pleasure. Pain, which is connected to terror or fear, falls under the biological drive of self-preservation, whereas enjoyment and pleasure attach to the human as a gendered and social being (Brøgger, Bukdahl, and Heinsen 1985: 7–8). Burke associates the sublime with self-preservation and pain, while he places the experience of beauty within the domain of community and society. Hence, the sublime, in contrast with beauty, is a fundamentally anti-social emotion. The sublime can only be experienced *alone*; sublime experiences are always subjective. On the contrary, beauty can be appreciated and celebrated in community with others.

Everything—in real life as in art—that provokes feelings of fear and terror Burke considers as foundational to the strongest possible emotion the human mind can endure, namely the sublime. Fear, in particular, can arouse in the conscious mind great ideas:

> Whatever is fitted in any sort to excite the ideas of pain, and danger, that is to say, whatever is in any sort terrible, or is conversant about terrible objects, or operates in a manner analogous to terror, is a source of the sublime; that is, it is productive of the strongest emotion, which the mind is capable of feeling.
>
> (Burke 1958: 39)

All general privations—emptiness, darkness, loneliness, and silence—are sublime, according to Burke, because they lead to a *fear that this is all there is*, and because they intensify the senses of the experiencing subject. Aside from these general privations, the feeling of infinitude or boundlessness can also be described as sublime, since when we are not able to grasp the totality or coherence of the world immediately, we are filled with fear (Burke 1958: 71–73). Fear is always caused by a force that is directly superior to the human experience. Human beings, Burke thought, never voluntarily submit to fear.

Sublime objects or phenomena create problems for the mind because they are too powerful, too great, too shapeless, or too obscure to be captured or comprehended right away. For this reason, they are intimidating to the human imagination. Human beings fancy the beautiful, as it can be understood and "captured" in an immediate way: it is accessible, easily decoded, and easily detected (and associated with the Pleasure of the Familiar). The sublime, on the other hand, creates terror, since it appears as being *greater* and *more than* the human.

Terror, horror, fear, and pain are thus, in Burke's understanding, crucial elements in the sublime experience. But just as importantly, sublime moods belong to the *imaginary* and are therefore not caused by real threats. We find here another reference to Aristotelian catharsis, which offers a release for the experiencing subject exactly because of its fictional character.

There is nothing sublime about threats to one's life (in the same way as bad design does not provide any kind of challenging aesthetic nourishment—more about this in Chapter 3, "The expression of flexible aesthetics" and Chapter 6, "The value of aesthetic sustainability"). It is the *representation* of danger, or perhaps simply a brief intervention in one's comfort zone, that propels the sublime experience, and that can lead the spectator or user towards an edifying mood of expansive consciousness and a sensorial surplus. It is the feeling of being torn away from everyday life with all its assumptions that starts the experiencing subject on her second- or minute-long *Bildung*, characterized by a movement from order to chaos to a new form of order.

It is important to note that the sublime is not solely characterized by negative emotions, such as terror, but also by pleasure—even a kind of pleasure that can transcend the more readily available pleasure caused by beauty. Nevertheless, for terror to lead to pleasure—or for the psychological or emotional chain reaction from pain to enjoyment, or from chaos (back) to order to take place—it is necessary to keep at arm's length the threat causing the terrifying emotion. Neither the threat nor the feeling of terror is *in itself* sublime. Purely from a position of safety—that is, through physical or artistic distance from the threatening object or situation—can terror appear pleasurable.

Burke's sublime experience consists of what can be termed a comfortable sort of terror, which exists in a middle position between imaginary death and intensified life. The culmination of the sublime experience is characterized by an intensification of the emotions that follow from the lifting of something negative; Burke terms this intensification "delight" (Burke 1958: 36). The delight that the sublime affords the experiencing subject, according to Burke, is fundamentally different from the kind of spontaneous enjoyment that comes from beauty, which he describes as a positive pleasure. The beautiful evokes an absolute and independent enjoyment, whereas sublime delight is derived from an external, negative source. Negative enjoyment, or comfortable terror, is associated with pain and the feeling of being close to death—that is, an *imaginary* threat to life. It is an essential part of experiencing "fear," but losing one's self is not part of the sublime experience. The sublime is close to ecstasy, but it involves by definition a "return home." Pure loss of self is destructive, and destruction has nothing to do with the aesthetic experience. On the contrary, the sublime aesthetic experience is edifying and self-defining.

In that magical moment when all-encompassing terror dissipates, the experiencing subject, Burke says, lets out a sigh of relief, which is accompanied by a sensorial, emotional, and corporeal intensification. The senses are no longer paralyzed by fear. The threat has abated, and the subject is left triumphant, clear-sighted, and balanced. An existential and temporal crisis has been averted in a moment of facing the shadowy side of life. It is for this reason that the subject comes away with a renewed sense of intensity and strength.

Individual encounters with the sublime are sure to be pivotal and insightful. The question becomes, however, whether the kind of elements or objects and phenomena that trigger the sublime aesthetic experience can be considered universal, or common to all of humankind. The following sections will make clear that the answer is yes, and it is subsequently relevant to include the sublime in constructing guidelines for the creation of objects and concepts that, in all likelihood, will challenge the recipient's aesthetic assumptions, prompting a potentially life-altering experience, linked as it is to the Pleasure of the Unfamiliar. The hope is that this will be an experience that creates a more durable bond between subject and object.

A more thorough discussion about this will follow later, but first I will expand further upon the intellectual history of the sublime aesthetic experience.

### Positive and negative pleasure

In his Third Critique, *Critique of the Power of Judgment*, the German philosopher Immanuel Kant (1724–1804) summarizes and synthesizes 18th century thought regarding the sublime and the beautiful. In the section on the aesthetic power of judgment, Kant, agreeing with Burke, makes a distinction between the beautiful and the sublime. Regardless, Kant also finds similarities between the beautiful and the sublime, as he considers both the feeling of the beautiful and the feeling of the sublime as being individual, subjective, and aesthetic forms of judgments, which in spite of their subjective nature are regarded as being *universally* valid. According to Kant, there exists a subjective but disinterested sphere that is common to all of humankind; in fact, human consciousness, to Kant, is structured, as included in the concept of *sensus communis*, according to the notion of subjective universality (Kant 2002: 161). The conditions of experience are thus independent of experience, *a priori*, or a predetermined frame. Any given object is shaped in the moment of experience according to the universal conditions of understanding.

The notion of subjective universality and the commonly valid aesthetic judgments are interesting in relation to the pursuit of the aesthetically durable—and, in the same breath, the universally appealing—expression. This theory can therefore help shape the foundation for developing universal guidelines in constructing a *durable* expression, or an expression that can provide everyone (or at least most everybody, or perhaps rather those who are open, or who have been educated, to being receptive to aesthetic experiences), regardless of cultural "baggage" and socially determined assumptions, with a challenging aesthetic experience.

Similarities exist between beautiful and sublime aesthetic experiences, but the basic difference between the beautiful and the sublime is nevertheless, in Kant's view, far more apparent than any similarities they might share. For example, beautiful pleasure and sublime pleasure are characterized by a difference in kind, since, while the pleasure evoked by the sublime is indirect and negative, the beautiful elicits a direct and positive enjoyment (Kant 2002: 129). The appreciation of beauty is peaceful. By contrast, the experience of the sublime moves or stirs us in a powerful way; the conscious mind is alternately attracted and repelled by the object cause of the sublime.

The dialectical movement that characterizes the sublime aesthetic experience is closely connected to the Pleasure of the Unfamiliar as well as the challenging (design) object encounter. In the same way, the peaceful, meditative, calm enjoyment characteristic of the beautiful is associated with the Pleasure of the Familiar as well as with the feeling of home and the tranquil sensation of belonging that accompanies it.

The Pleasure of the Unfamiliar is characterized by an almost magnetic and corporeal, sensuous kind of attraction (the fingers just *have* to touch, investigate, and prod) as well as a bodily and psychological *taking-a-step-back-from* to recover control through a moment's reflection and a processing of one's sense impressions. Herein lies the strength of this form of aesthetic pleasure; the fact that what overwhelms or disgusts us is that much harder to "shake off." The feeling of being in a room with a challenging (design) object lingers, so to speak, for a while afterwards and it can be difficult to forget the experience.

The distinction between positive and negative pleasure in Kant can be traced back to Burke. *Delight*, in Burke's thought, connotes negative pleasure and is linked to the sublime, whereas *pleasure* belongs to the beautiful. However, the central difference between the beautiful and the sublime, according to Kant, is the fact that (natural) objects can be considered "beautiful," but never "sublime." Beauty desires form, while the sublime is formless, bounding ever outward. The sublime is absolutely great: "It is a magnitude that is equal only to itself" (Kant 2002: 134).

The feeling of beauty is a result of experiencing form and, hence, a limitation. As such, it is connected with the experience or belief that forms can be *under-stood*, comprehended, and categorized. By contrast, the sublime is characterized by formlessness—or by an experience of forms that seem so vast or abstract that they cannot immediately be apprehended (even less so, comprehended). In this way, it belongs to *reason*, meaning the ability to juxtapose and to synthesize through the power of reflection. The sublime can only be acknowledged through speculative reason (see Lyotard 1991: 137). The Kantian sublime, thus, does not have much to do with corporeal, phenomenological activity and experience; rather, this kind of experience should be considered reflexive.

Kant distinguishes between the *mathematically sublime*, which deals with the effect of the absolutely (physical) great on human consciousness, and the *dynamically sublime*, which concerns the intimidating powers of nature. Both forms of the sublime are connected with human cognition, or reflection (Crowther 1989: 85–86). The one is in no way subordinate to the other. Kant's structuring of the two should not lead us to think that there are two forms of the sublime as such. Nevertheless, this way of distinguishing between the two is interesting for reasons that will become clear in the following paragraphs.

The experience of the *mathematically sublime* is related to a kind of mental exhaustion that happens when consciousness is confronted and challenged by a physical object of extreme vastness or apparent infinite dimensions. Kant gives as examples appearing in front of an Egyptian pyramid or the St. Peter's Basilica in the Vatican. In this way, he demonstrates how, when viewed from an *appropriate* distance, such worldly grand monuments, which can otherwise be measured mathematically, transcend the measurable and appear *absolutely great*. From an appropriate distance, the different parts of the monuments cannot be juxtaposed, and, consequently, it is impossible to form a complete impression of the whole. In other words, an object can appear formless as it overwhelms the senses of the subject.

I can definitely relate to the thought that, at a certain distance, monuments, boulevards, or cityscapes might seem inherently *expansive* as they merge with the surroundings, perhaps especially when combined with certain weather conditions. On a personal level, when visiting the ancient city of Petra in Jordan, the combination of whirling dust and the magnitude of the ruins created an overwhelming sense of grandeur. I've had similar experiences looking out over the Champs-Élysées during a heavy rain shower or when approaching the harbor of Essaouira shrouded in heat haze and sea fog, accompanied by the fluttering wings of seagulls. It is namely the overwhelming confluence of impressions, as described by Kant, that characterizes such grandiose experiences; however, sensuous embodiment— the smells, the feeling of wind and rain against my face, or the hot dust—is equally important to consider when discussing the sublime, I argue.

As part of the *dynamical* sublime experience, the subject undergoes a triumphant feeling of mental superiority. When the subject, from a position of safety, is confronted with an overwhelming, dynamical, and threatening phenomenon (e.g., the storm passing over the mountains, according to Burke's representation of the sublime), she is challenged. But the challenge leads to a resistance: the power of the imagination *imagines* that it is forming a point of resistance against destruction. This hypothetical situation, so to speak, makes the human subject *spiritually* and mentally stronger. Coming face-to-face with the dynamical and potentially destructive phenomenon, the subject's mind generates a resistance that makes her feel strong and superior (Crowther 1989: 148).

In the sublime moment, which is the culmination of the sublime aesthetic experience, the experiencing subject is filled with a special kind of raw stamina because it has managed to surpass the feeling of danger or of not being able to cope with the phenomenon facing it. The sublime experience thus involves an experience of self. It is the self that the human subject is confronted with in the sublime moment as insight into one's own responses; expectations or connotations; basic assumptions; weaknesses; and, finally, one's own grandeur and mental capacity.

But, as an experience of self, is the aesthetic experience solely of the mind, or is it rather characterized by an interaction between an object and the individual? In the case of the latter, might the aesthetic experience better be described as a *sensuous* experience rather than as a reflexive one? And could we then characterize an object *in-itself* as being sublime? The following sections seek to answer these questions.

## The stages of the sublime

> The act of contemplation, for a human being, is a both/and situation: to contemplate something and, at the same time, experience oneself contemplating. The moment of contemplation is therefore also an existential situation. It is to discover one's own position by way of an object and to discover an object's position by way of one's self.
>
> (Ørskov 1999: 26, transl.)

This quote provides an appropriate frame for discussing the different stages of the *sublime* aesthetic experience. As part of the aesthetic experience, the individual discovers knowledge about the self as well as about the object or phenomenon experienced. The aesthetic experience involves a measure of self-knowledge, as it helps the subject frame its own expectations about the world and objects. Furthermore, included in the sublime aesthetic experience and the Pleasure of the Unfamiliar is the possibility of expanding one's horizon.

Many thinkers have characterized the sublime as being ineffable, as the sublime aesthetic experience does not lend itself easily to words. To a large degree, it consists of an agitation of the emotional and mental life of the subject, but, at the same time, it is highly sensuous, involving a close attachment to the object or phenomenon that has triggered the experience. The triggering object or phenomenon is relevant in regard to the concept of aesthetic sustainability, as it relates to the sublime or the Pleasure of the Unfamiliar. This kind of object or phenomenon contains a large measure of durability, since the subject encountering it will be able to spend a great deal of time contemplating, sensing, and experiencing it.

The following quotation from Karl-Ove Knausgaard's first volume of *My Struggle* (2012) describes how it is possible, time and again, to draw aesthetic nourishment from an object (in this case, from a painting) in order to set the mind in motion:

> I sat leafing through the Constable book for almost an hour. I kept flicking back to the picture of the greenish clouds, every time it called forth the same emotions in me. It was as if two different forms of reflection rose and fell in my consciousness, one with its thoughts and reasoning, the other with its feelings and impressions, which, even though they were juxtaposed, excluded each other's insights. It was a fantastic picture, it filled me with all the feelings that fantastic pictures do, but when I had to explain why, what constituted the "fantastic," I was at a loss to do so. The picture made my insides tremble, but for what? The picture filled me with longing, but for what? There were plenty of clouds around. There were plenty of colors around. There were enough particular historical moments. There were also plenty of combinations of all three. Contemporary art, in other words, the art which in principle ought to be of relevance to me, did not consider the feelings a work of art generated as valuable. Feelings were of inferior value, or perhaps even an undesirable by-product, a kind of waste product, or at best, malleable material, open to manipulation. Naturalistic depictions of reality had no value either, but were viewed as naïve and a stage of development that had been superseded long ago. There was not much meaning left in that. But the moment I focused my gaze on the painting again all my reasoning vanished in the surge of energy and beauty that arose in me. *Yes, yes, yes*, I heard. *That's where it is. That's where I have to go.* But what was it I had said yes to? Where was it I had to go?
> (Knausgaard 2012: 286–87)

The painting that Knausgaard is here contemplating, and which he keeps returning to for emotional fulfillment, overrides all reasoning or, rather, all filters. It is an expressionistic landscape painting and, being a rather traditional specimen, it is not particularly interesting to a late-modern audience. In no way is it critical, provocative, or innovative; regardless, it makes Knausgaard's "insides tremble," and it is impossible for him to explain why. The sublime, ineffable beauty of the painting reduces him to an exposed, vulnerable, and receptive state.

Burke is considered an empiricist or a sensualist, which is why, for him, an experience of the sublime is closely connected with the sublime "object" and the subject's impression of it. (Object is here placed in quotation marks, as Burke, in a fashion typical of his period, associates the sublime with nature and its grand, overwhelming phenomena, and not "objects" as such). In his *Enquiry*, Burke sets out to determine a number of psychological factors that might explain why human beings are moved by the immensity and formlessness of sublime objects. In other words, he focuses on the subject-effect of the sublime. However, he does not see the sublime as being solely an effect of individual consciousness. For Burke, and for most other 18th century English thinkers, the sublime is bound to a statement about the nature of the stimulating, sensory object. This can be related to aesthetic sustainability, and more specifically to the experience of things and their potentially magical or auratic effects. Chapter 5, "The magical thing," will deal with this in more detail.

In contrast to Burke, Kant theorizes that the sublime implicitly contains a statement about the nature of *sensory consciousness*—and *not* that of the sensory object—thereby associating the sublime with the subject's consciousness *alone*. In this way, Kant transforms the sublime, as an aesthetic concept, into a purely subjective category; it is not a quality that inheres in the object, but it is rather a state of mind that is triggered by the object. (There *is* a *triggering* object or phenomenon, but this object and its quality are not essential to the sublime experience). The grandeur of the sublime is thus not empirical, but depends instead on an idea or intuition. The sublime does not spring from the object, but the subject—that is, the subject's reason, not its senses. And the experience of the sublime is fundamentally bound to the human ability to rise above what is threatening or challenging. Put differently, Kant leaves behind Burke's empiricism to create instead a transcendental philosophy, which will be expanded upon in the following paragraphs.

Kant's analysis of the sublime can be called *negative*, in the sense that it introduces an *object-less aesthetic*. The aesthetic experience of the sublime is not, by necessity, founded in the object. Immense, uncontrollable nature can activate the sublime experience, as it challenges human imagination and "forces" it to contain grandness and intensity. But nature cannot *as such* or *in itself* be said to be sublime. The sublime, as an aesthetic concept, is thus entirely subjective—it is a state of mind, not a quality of the object, even though it might serve as the activating force of the sublime experience.

According to Burke, the culmination of the sublime experience must be characterized as intensive, thrilling, and joyous terror: "the delightful horror." Kant, on the contrary, sees the sublime experience as concluding in an exalted calm. The sublime experience culminates with the subject's realization of its own ability to rationalize and thereby simplify and understand. The realization follows that, despite external disturbances, the human possesses a firm guiding principle, namely reason. As the subject comes face-to-face with the power or seeming boundlessness of nature, the former will create conscious ideas of comparable magnitude, exactly because the human has the ability to reflect upon and define its own experiences. The joy that Burke considers the fulcrum of the sublime is for Kant merely a step towards exaltation, which is characterized by a feeling of freedom and wholeness.

Experiencing the sublime, according to Kant, in ways reduces human attachment to worldly phenomena or life. The sublime, in Kant, corresponds to the development of an inner strength or calm, rather than an intensification of the senses and the joy of life; in experiencing the sublime, the subject, in an almost meditative way, turns inward, away from the physical world. The sublime is *in* the human; it should not be sought in natural phenomena, but in the world of ideas: "true sublimity must be sought only in the mind of the one who judges, not in the object in nature, the judging *of* which occasions this disposition in it" (Kant 2002: 139).

The sublime aesthetic experience, inspired by Burke and Kant, is characterized by both sensuousness and reflexivity. In the sublime moment, the subject is at once present, *corporeally*, but also able to reflect and expand on its own productive capacity and horizon as a human being. However, an object *in itself* can possess *auratic* or *magical* qualities that render probable the sublime aesthetic experience. But it is the interplay between object and subject (the very *experience* of the sublime) that is truly interesting, as it is here that the aesthetic horizon-expanding experience can take place.

The sublime aesthetic experience consists of a few different "stages." Inspired by Kant's ruminations in *Observations on the Feeling of the Beautiful and the Sublime* (1764), the aesthetic experience can be divided into three stages—and these three stages can conveniently be transposed onto the design experience:

1   The initial "meeting" of subject and object/phenomenon (an encounter that can be both sensuously and psychologically challenging).
2   The subject's attempt at understanding the greatness of the object/phenomenon, "capturing" or comprehending its formlessness to make sense of its asymmetrical nature or its confusing signals.
3   The reason makes clear to the imagination what is going on, as, for example: "the reason why you don't immediately understand what is happening is because you are experiencing an explosion of forms or a chaotic composition that you are not used to being confronted with, or that you have not previously encountered." The imagination will then produce a measured response, allowing the subject to return to the world with a calmer and more present mindset as well as an expanded horizon and understanding about its surroundings. The kind of expanded consciousness produced by encountering the sublime is in part cognitive but also highly sensuous; an intense feeling of presence spreads to each pore and every nerve ending of the body, and what follows is a sensorial explosion of complete bodily awareness.

In the realm of design strategy, stage 2 can suitably be extended. This stage directly challenges the understanding, or imagination, which, at this point in the aesthetic experience, struggles to synthesize and comprehend what is going on. A break is imminent. This stage might only last for a moment, after which the understanding can comprehend the situation and the subject will experience a payoff in interacting with the object. Of course, the stage can be prolonged, design-wise, rendering the experience more complex.

In terms of design, it takes a thorough understanding of one's target audience in order to know how long it is "acceptable" to expose them to chaos. Are they seeking a challenging experience, and will they feel *aesthetically nourished* by a moment's discomfiture? Or, contrarily, will they be more aesthetically satisfied by being able to quickly decode and use the design object? It is by carefully matching a product to the aesthetic needs of the audience that it becomes aesthetically sustainable.

The sublime aesthetic experience is similar to the Pleasure of the Unfamiliar. It is a kind of pleasure that is triggered by the idea that (put simplistically), "A shoe can *actually* look *like this*"; or, "A chair can *also* feel *this way*"; or, "A drawer can *likewise* open in *this way*"; or, "A dress can *evidently* have *this shape*". Experiencing a sublime aesthetic object leads to a new understanding of forms and materials as well as how they shape our perception of the world.

In other words, the Pleasure of the Unfamiliar contains an instructive quality; it adds to one's experiential knowledge, and it might thus prove helpful when faced with a different kind of kitchen drawer, for example, or a piece of clothing that is not easily decoded or detected. The previous sublime experience has expanded one's horizon, if only by a degree. In the sublime experience, the subject's connotative framework comes into focus. The subjective framework is forced to expand, so to speak, when encountering an object that falls outside its parameters of understanding; the result is a renewed self-knowledge and an insight into the world that expands one's horizon.

Due to the challenging nature of expanding the subject's basic assumptions and connotative framework by involving the sublime, and the Pleasure of the Unfamiliar, this type of aesthetic experience can be considered sustainable or durable. Human beings need to be challenged and to feel that they are "moving" in some way. Durability, or sustainability, is not just about maintaining the status quo or preserving what is familiar. Sustainability emanates with the possibility of renewal and self-development as well as self-stabilization. The aesthetically sustainable object, in this regard, can be described as the type of object one never tires of contemplating, touching, or exploring—because of, rather than in spite of, its complexity.

The Pleasure of the Unfamiliar is a "troublesome" form of aesthetic pleasure, understood in the way that it is neither easily accessible nor immediately comfortable. It challenges and questions everything familiar, and it pushes at the boundaries of subjective consciousness. But this is exactly its strength. This kind of pleasure can create a sustainable bond between subject and object that is founded on a pivotal aesthetic experience, which is considerably less insignificant than all the other sense impressions bombarding us on a daily basis in today's late-modern world. The Pleasure of the Unfamiliar stays with the subject long after the aesthetic experience itself has ended.

## Breaking with universal aesthetic principles

The beautiful and the Pleasure of the Familiar involve an experience of impeccability, harmony, symmetry, and limitation, as well as an experience of being able to easily comprehend or decode and detect a given object. By contrast, the sublime and the Pleasure of the Unfamiliar demand the dissolution of forms, harmonies, and symmetries; this side of the aesthetic spectrum can be accessed through objects or concepts that are difficult to detect and decode, not easily "taken in."

In the section on "Adhering to Universal Aesthetic Principles" in Chapter 1, I described what can be termed universal preferences for what serves as the most pleasing expression or idiom. For despite the fact that people are different in many ways, physiologically we are all nearly the same. Because we have the same type of limbs, we tend to experience space in the same way; likewise, our taste buds work the same way, which is why it is possible to describe food in terms of sweetness, acidity, etc. Consequently, it must be possible to construct guidelines for how shapes and colors are perceived and processed by the senses as well as how, *a priori* (prior to cultural and social connotations being internalized by the individual), immediately balanced and easily decoded expressions are grasped. Following this, it should be possible to work out guidelines for how to most effectively break the individual's comfort zone, thereby eliciting the kind of aesthetic experience that is characterized by the Pleasure of the Unfamiliar.

However, before attempting to challenge the recipient in a design context, it is imperative to know and understand the universal aesthetic principles. Without knowing the "rules", it is nearly impossible to break them in any intentional way. Breaking the universal aesthetic principles will inherently challenge the recipient, and her basic assumptions about the world will be torn apart, if only for a moment. As a consequence, the recipient will leave the encounter with new or distorted assumptions and with the experience of finding pleasure or aesthetic comfort in a wholly different guise than usual, along the lines of the sublime aesthetic experience. In other words, unfamiliar aesthetic experiences disrupt not only notions of beauty, comfort, and attraction, but also the recipient's worldview, leading to new and pivotal insights.

According to French philosopher Jean-Francois Lyotard (1924–1998), the sublime aesthetic experience (which he considers essential to the development of human consciousness) is important for the seemingly basic reason that *something happens*. When shocked by unusual combinations, asymmetrical compositions, or chaotic structures, the human subject, more or less forcefully, is pulled out of daily routines and away from dominant expectations or basic assumptions. Suddenly, one becomes more present and attentive The sublime *something* happens right in front of one, and it is a struggle to apprehend and comprehend it:

> The arts, whatever their materials, pressed forward by the aesthetics of the sublime in search of intense effects, can and must give up the imitation of models that are merely beautiful and try out surprising, strange, shocking combinations. Shock is, *par excellence,* the evidence of (something) *happening,* rather than nothing, suspended privation.
>
> (Lyotard 1991: 100)

Contemplating objects with what could be called a "beginner's mind" (an unprecedented state of mind) entails a movement away from one's comfort zone. It is this mindset, brought about by a sublime experience, that can lead to an expansion of one's horizon.

Human beings constantly seek to organize and structure the many different impressions that belong to an average day. The reader will remember from the section on the Pleasure of the Familiar that it is immediately easier to detect or decode objects we are used to or that possess a certain symmetry or inherent clarity of structure, which aid the eye and the imagination to make sense of them. By contrast, chaotic structures and unusual or unknown combinations strain the eye, so to speak. The eye flits across the unruly surface of the object in an effort to organize and order, as quickly as possible, the confusing material it is faced with. It is part of human nature to want to understand, organize, and structure any and all sense impressions, no matter how confounding they may be.

In Kant's view, geometrical shapes are too perfect to elicit an aesthetic experience. Insofar as they cohere with the underlying concept or idea—thus possessing the *precision* that the ancient Greeks sought and celebrated—geometrical shapes can be grasped, but they do not evoke emotion, and, most importantly, they do not move the imagination to free and new (mental) lengths.

Forms or phenomena, on the contrary, that posses a degree of immeasurability, or that do not appear constrained, stimulate the human imagination (Böhme 2010: 26)—hence their ability to induce a sublime aesthetic experience. The pleasure associated with experiencing immeasurable objects—indefinable or formless objects—can be defined as enjoying one's own emotional and mental activity. Namely, the pleasure consists of being challenged and struggling to understand and decode the phenomenon present to view. Furthermore, part of the pleasure comes from having one's comfort zone (momentarily) breached.

Lyotard considers the aesthetic of sublimity foundational to 20th century avant-garde art; the sublime is an essential element of the artists who revolted against the Romantic "closed" pictorial space as well as against the attempt at creating, through the artwork, a wholeness not possessed by the real world. The aesthetic of sublimity rather strives for the indeterminate and unproductive. It liberates pictorial art from the demand to facilitate concrete messages, thus setting free the power of creation (Brøgger, Bukdahl, and Heinsen 1985: 13). Similar to Kant, Lyotard distinguishes between two forms of sublimity: first, the *nostalgic* or *melancholic* sublime, which strives for unity with nature, the universe, absolute spirit, or the divine; Romantic poets like William Wordsworth (1770–1850) are exemplary of this mode. Mixing joy and pain, the second form focuses on intense, aesthetic experiences of the *now*. Whereas Lyotard associates modernity with the nostalgic, Romantic sublime, he relates postmodernity to the other form of sublimity, which, to him, encompasses the *true* sublime—or at least the kind of sublimity he finds the most interesting.

The (postmodern) intense sublime moment revolves around the pleasure of *It is happening*, the delight of *something shocking* (an object or phenomenon) as well as the fear that *It will stop happening*, and that nothing further will happen (see Lyotard 1991: 99–100). The dialectic of delight and fear, pleasure and pain, or the *It is happening* and the *It will stop happening* characterizes Burke's definition of the sublime as well. In Burke's thinking, the sublime is associated with terror caused by privation: darkness (privation of light); loneliness (privation of company); silence (privation of language); emptiness (privation of presence); and, finally, death (privation of life). The sublime deprivations are related to a central terror: "What is terrifying is that the *It happens that* does not happen, that it stops happening" (Lyotard 1991: 99).

Nevertheless, this terror, as mentioned, is also related to the *delight* of *It is happening*. As part of the postmodern sublime, the individual senses intensify. Lyotard, similar to Kant, locates the sublime in individual reason; again, the sublime is a state of mind, not a quality of the object, which might of course initiate the sublime experience. The sublime experience is one of pure presence, and the subject is propelled into an extreme *here-and-now*.

According to Lyotard, this is the purpose of all art; through surprising, unusual, and shocking combinations, the viewer is thrown into *It is happening!* The pleasure derived from the sublime experience depends on being completely present in the *now* of the artwork—for possibly quite a brief moment. Herein lies the potential of new insight and an individual experience of the immediate world of the subject; further, this experience contains a sustainable or durable component. Following the satisfying outcome of the sublime experience, the subject will (hopefully) be amenable to seeking out future eye-opening sublime aesthetic experiences, as these contain a particular kind of aesthetic nourishment. Lyotard raises a number of points regarding the sublime that can adroitly be adjusted to fit the design experience and the role of the designer.

As mentioned previously, the human eye is automatically drawn toward symmetrical patterns. Because of its organizational and structural nature, the eye can be made to lose its composure, so to speak, by not being given what it wants. Intentionally turning things on their head, either figuratively or literally, it is possible as the sender of an object (or concept) to momentarily confuse the recipient's senses, thereby forcing her to stop and think or to examine the object more closely.

Disrupting the harmony of color,[3] for instance, can lead to a contemplative "break" in the humdrum existence of everyday life. By challenging the contrast of extension, allowing the visually more "expansive" color to dominate over the more subtle hues—which would usually take up more of the canvas in a harmonious composition—the eye of the beholder can be made to wander and to return, over and over, to the dominant color in a dialectical movement.

In a composition, breaking with symmetry and the golden section are additional ways of forcing the eye into high tension with the intention of producing a psychological interval; this interval, typically of short duration, has the potential to upset the viewer's expectations and connotations—that is, until reason has had a chance to reassure the imagination.[4] In the same way, using ostensibly inert (or otherwise nontraditional or unconventional) materials in regard to the form, function, or aesthetic of the object can produce the subjective *interval* (stage 2 of the sublime aesthetic experience), compelling the recipient to investigate, consider, and examine her mind for what is happening. This interval is productive and important because it either challenges or disrupts the subject's comfort zone; it may even become an element of the subject's thoughts and senses for some time after the experience (including the time of decoding or detecting) has ended.

In the remainder of the chapter, I will occasionally come back to how designers can plan the interval—or expand stage 2 of the sublime aesthetic experience—in order to create the potential of a sustainable bond between subject and object.

## Prolonged decoding

That something which is either difficult to decode or difficult to understand should be able to elicit any kind of pleasure, on the face of it, seems inherently contradictory. Nevertheless, the Pleasure of the Unfamiliar exactly involves the experience of not immediately being able to understand, grasp, or decode an object (or concept or phenomenon). The Pleasure of the Unfamiliar comprises an aesthetic experience of being faced with *something* ineffable. That *something* might seem similar to a previously encountered object: the form might be recognizable and pleasing, to an extent; but there is something about it that makes it difficult to place within one of the usual sensuous categories at one's disposal. As previously stated, it might be that the shape, material, or color combination is different than anything experienced on an earlier occasion—or it might be that the recognizable element does not fit the current object because it breaks with one or more aesthetic conventions.

However, it could also be that the experience in question would have more far-reaching implications; for example, it might not be possible right away to determine the nature of the object one is dealing with. (Is it a kitchen appliance or a piece of craft art? Is a jacket or a pair of pants?) Any such ambiguous object requires a certain amount of time to decode and, subsequently, to place within one of the psychological categories used to organize, structure, and understand the world. Or, perhaps, the present object demands an entirely new category be created. Alternatively, a category of *sampled* or "hybrid" objects (as, for example, a short-sleeved coat made from furniture upholstery; or a sofa made from both soft and hard elements, and that is as low as a madras) is required.

I have previously described—and illustrated—how to "conjugate" the sublime aesthetic experience,[5] the idea being that the aesthetic sublime experience can be more or less complex, which is why, conceptually, it can conveniently be converted to the category of prolonged decoding (with the easily decodable as its opposite), illustrated in Figure 2.

Easily decodable

Prolonged decoding

*Figure 2* Prolonged decoding versus easily decodable

Designers can seek to provide audiences with an experience of something that is either easy or difficult to decode—placing different objects on extreme ends of the aesthetic spectrum or somewhere in the middle. The placement will ultimately be decided by the intention of the object (or concept), by the design category in question, or by the audience or segment for which one is designing. In some contexts, prolonging the decoding time substantially could be highly effective in terms of momentarily disorienting the recipient; in other cases, slightly confusing the recipient's notion of what is usual might be preferable.

For a clothing designer, for instance, aiming to create durable pieces or outfits for an audience-type who likes being challenged, who prefers garments that break with the norm of what clothing is and can do, and who likes to stand out, increasing the confusion of signals (and thus prolonging the decoding time) is appropriate. In such a case, the choice might fall on producing a modular garment system that involves the user in combining different clothing elements, thereby giving the design an uncontrollable and anarchic feel. As a different example, a concept aimed at minimizing the amount we are used to washing our clothes in the western world is likely to challenge traditional notions of what it means to dress well and what "nice" clothes look like. These two examples are by design rather simplistic, but the point should be clear.

If the audience or segment in question, however, does not typically want to stand out or go against the grain, but instead finds comfort and pleasure in belonging and blending in, this does not automatically mean that one cannot choose to work with a product that prolongs the decoding time (in order to catch the attention of the user, for example, or to create a sustainable link between subject and object through a challenging aesthetic experience). In this case, one should take care to use fewer unrecognizable elements and confusing signals. Designing a reversible coat is an example, which by virtue of its aesthetic and functional flexibility can be used in different social contexts, rendering it potentially more sustainable. This way of approaching the creation of durable clothes is much easier to decode, and the added element of unfamiliarity is not rooted in any immediate confusion over what kind of garment we are dealing with (it is obviously a coat, the style of which might even appear rather traditional); the time of decoding, in this instance, is much more likely to be prolonged by the reversible feature, a potential cause of momentary confusion, but which does not take long to decode. The *payoff* is bound to arrive relatively quickly in this case, and the design is thus located fairly close to the middle of the aesthetic scale (Figure 2), drawing near to the easily decodable and the Pleasure of the Familiar.

The horizon of the experiencing subject is expanded as part of the prolonged decoding time of the sublime aesthetic experience, which engenders a new, "extensive" (aesthetic, sensuous) category of experience—in the sense that the subject commands additional forms or expressions in the categories of "chairs" or "coats," for example, as an outcome of the experience in question. It might also be that the experience itself creates for the subject an entirely new category of objects: garments that do not need washing, for instance. Moreover, objects or concepts of prolonged decoding force the conscious mind to struggle, and thus to expand, to grasp the phenomenon facing it.

Philosophically speaking, this kind of horizon-expanding experience leads to a new form of insight; the familiar is challenged, but the split second when the *payoff* arrives ("Ah, I see, the coat is reversible"; or, "This drawer opens in a much different way"; or, "I've never seen this kind of surface material on a cup before") is similar to discovering a new insight about one's life: *something* which forms a "before" and "after"; *something* pivotal. This comparison might seem somewhat hyperbolic, but it is to be understood in the following way: when next faced with a coat, a drawer, or a cup, one's previous, insightful experience will influence the experience of the new object; here in lies the "before" and "after."

The challenging aesthetic experience, which *expands* the conscious mind, and which can seem uncomfortable because it requires a different kind of focus, presence, and concentration (experienced as uncomfortable because we are used to knowing exactly how to handle the objects of our surroundings), can therefore remind of pivotal, self-enlarging experiences in one's personal life. The self is thrown into the deep end, but by managing to get out, and in the moment when reason kicks in, the experience becomes one of exhilarating insight. The scale is perhaps of a different sort in the context of design objects, but the feeling is nonetheless similar.

By no means, however, is aesthetic value confined to objects or concepts of prolonged decoding; the easily decodable can contain immense value as well.[6] Both kinds of sensuous, stimulating encounters with the world and the objects in it can have an edifying, or even an educational, effect, as they "teach" the subject how to be in the world. This should be understood in relation to how, when encountering objects that either meet or challenge one's usual expectations, either one's basic assumptions are affirmed (as in, "This drawer opens and closes exactly as I thought it would") or one's horizon is expanded (as in, "Ah, this coat is reversible"). Such experiences, which might seem insignificant and ordinary, are crucial for the individual to feel comfortable and "at home" in the world. Following an edifying object interaction, the individual will face the world either with the comfortable, affirming feeling of being able to understand and decode immediately her surroundings or with the knowledge of having had her horizon expanded through interacting with and decoding an object that appeared incomprehensible and alien. Additionally, the affirming as well as the challenging design experience contains pleasure: the Pleasure of the Familiar and the Pleasure of the Unfamiliar, respectively. Both forms of pleasure are aesthetically nourishing.

As described and illustrated previously, the aesthetic experience associated with prolonged decoding can be "conjugated" (Figure 2). As a designer or sender of a product, one can choose to challenge the recipient to a lesser or greater extent. As mentioned, the decision should be based on what makes the most sense in terms of the product and in terms of the audience type. Importantly, as related in the section on the sublime, the recipient or audience *must* be able to decode or understand the object (or concept) eventually—otherwise the design experience will be a poor one.

Facilitating the act of decoding—as intended by the designer, that is—requires a detailed knowledge of the recipient. Knowing the recipient and her frame of reference is crucial to the design process and to the marketing planning. It is just as crucial, however, to know what makes sense in relation to the product one is aiming to create. Is it a highly functional product (as for example a kitchen aid or a rain suit)? Or is it rather based on identification? Typically, one would choose to incorporate confusing signals and prolonged decoding into products that are meant to *boost* or subtend the recipient's sense of identity. Products of identification can range in type from furniture to bicycles to clothing—products that are all supposed to function in a certain way (sitting machines, transportation machines, keeping-warm-and-dry machines), but that, in our culture, are also greatly supportive of the recipient's identity and lifestyle. In creating these kinds of products—which also strive to be durable and sustainable—experimenting with prolonged decoding is a viable option. This element of the aesthetic experience can elicit an immediate sense of curiosity and interest, but also it can lead to the foundation of a sustainable bond between subject and object, a bond conditioned by fascination, interaction, or flexibility, for instance. Prolonged decoding is often expressed by a certain degree of objective flexibility, something which I will return to in the following chapter about aesthetic flexibility.

Phenomenologically, the Pleasure of the Unfamiliar has to do with challenging the subject in a sensorial and corporeal manner. This can be achieved, for instance, by using materials that feel different than they appear: the eye might have formed an expectation, perhaps, that the object will be hard, heavy, or cold to the touch—but as the hands begin exploring it, and as the tactile sense switches the mind to *stand-by*, the object feels soft, light, and warm instead. In a different scenario, the surveying eye and the experiencing body can be challenged by creating asymmetrical or ostensibly inharmonic compositions and shapes. In this way, the subject will learn something new about the world, leading to an insight of the senses or a corporeal being-in-the-world, rather than one of thought. The design experience can help the subject reach a corporeal or sensorial form of insight, which is prior to reflection, and which contains a potential source of knowledge about what it means to feel at home in the world, and about how to be present in the *now*.

According to phenomenologists such as Merleau-Ponty, consciousness and body are one: only through a bodily being-in-the-world can the human being become of one self. Bombarding the senses by using, for example, unfamiliar tactical elements can open up passages for the subject to previous experiences, thereby leading to an insight about something that used to be, or is, crucial to an understanding of self. It is essential, nevertheless, that the "step" preceding any reflexive insight is founded on bodily presence.

Most of us probably know to identify the arrival of spring or summer by the smell of flowers, grass, wood burning, ocean spray, and warm wet asphalt; these smells can transport us back to the realm of childhood or youth in a split second. The kind of intense presence marked by this sensuous experience, which at the same time acts as a portal into earlier times and experiences, is characteristic of insightful existence. Past and present become one in such moments of being-in-the-world, which are sources of insight into the workings and connections of existence.

In like manner, experiencing the tactility of a design object's surface can lead to a momentary merging of past and present. When encountering a challenging object that turns one's expectations upside down, elements of "homeliness," in the form of familiar structures or materials,[7] help create a balance between the known and the unknown. This kind of balance prepares the self to venture deeper into the Pleasure of the Unfamiliar and to experience moments of prolonged decoding.

In Chapter 4, "Designing the Temporal Object," and Chapter 5, "The Magical Thing," I will return to the insightful thing-experience.

From a semiotic perspective, designers can work with intentionally *charging* the design object with surprising moments and ambiguous signals, thus prolonging the time it takes the recipient to decode the object. Products can be coded in such a way as to make the decoding process comparatively difficult for the recipient— for example, by attaching a contradictory linguistic message to the object or concept. A clothing collection or series of interior designs might be given a title that is not immediately obvious to recipients, because it appears surreal or ambiguous, or perhaps because it consists of a compound word or term of one's own invention. It might also be that an unconventional onomatopoetic word, such as *kapow* or *shhh*, has been used in the marketing material or in the emblem of the collection in an effort to anchor the sound in the recipient's consciousness. In this way, it is possible to associate playful or even childish elements with a perhaps more serious or adult design universe—and it is possible that doing so captures a different mood for the recipient than what would otherwise be expected.

The Pleasure of the Unfamiliar, to a large extent, is about making one's recipient or audience pause for a moment; and it is about turning things on their head, either literally or metaphorically.

In addition to linguistic means, the prolonging of the decoding process can come about through an ambiguous or unconventional juxtaposition of visual elements. Creating a clothing collection, for instance, consisting of a collage-like juxtaposition of sampled styles and historical references could be one way of intentionally confusing the recipient. In such a case, the experiencing party would have to put in a significant effort in order to understand the underlying conceptual idea behind the collection; to help recipients navigate the collection, a less ambiguous title would be preferable. Titles function as anchoring points, as they ensure that certain (intentional) connotations remain, while other, unintended, meanings are left to pass on, unattached to the collection.

As previously mentioned, it is crucial to the aesthetic experience to be able to reach a kind of understanding or decoded meaning. The penny must be allowed to drop, so to speak, before a *payoff* experience becomes possible. If not, the Pleasure of the Unfamiliar will remain unfulfilled, and the recipient will not reach the satisfying culmination of the experience, which would otherwise become part of her continued frame of reference. An unrealized object or concept will typically be considered as bad design.

The *payoff* can be secured by knowing one's audience and their frame of reference, allowing one to work towards either accommodating or challenging it. In part by using connotative anchoring points, one should try to eliminate any possible misunderstandings. Desiring to create a visually highly obscure expression, as in the example above, that includes many different signals and elements, one should make sure to signpost the path for the audience in advance. Anchoring potentially "flighty" visual elements can happen by using a title that names the main reference points of the "collage." In other words, it is not advisable to increase the level of confusion by attaching a neologistic compound word to an already visually daunting or ambiguous expression.

Prolonged decoding is not the same as impossible decoding. Experiencing a moment of prolonged decoding is meant to be pleasurable: a brief interval encouraging someone to stop and wonder, or struggle to understand something, that has immense value in terms of expanding her horizon and "exercising" her powers of reason.

## Conclusion

The Pleasure of the Unfamiliar is closely related to the sublime aesthetic experience, which is inherently, and crucially, challenging. But a question remains: What does it take to challenge the late-modern human being, who is used to being bombarded with intense and dense sense impressions, as well as with an overwhelming amount of insignificant and even disturbing information? Pushing the boundaries of one's contemporaries—who are used to navigating vast urban areas and encountering a multitude of faces, objects, sounds, and smells—might seem an insurmountable endeavor. But perhaps this is not the case.

> I went into the street with the cup in my hand. A slight feeling of unease arose within me at seeing it out here, the cup belonged indoors, not outdoors; outdoors, there was something naked and exposed about it, and as I crossed the street I decided to buy a coffee at the 7-Eleven the following morning, and use their cup, made of cardboard, designed for outdoor use, from then on.
>
> (Knausgaard, 2012: 263)

This quotation from Knausgaard's *My Struggle* precisely proves my point. In reality, it does not take that much to startle or challenge us. By nature, human beings are creatures of habit, not least of all in a physical or sensorial sense. When something feels wrong or different, we tend to react instinctively with physical discomfort. Stored deep within us is a tendency to want to structure and categorize our surroundings to make them easier to comprehend and move through. Such structuring can be deeply satisfying, of course, and it is this tendency, or need, that the Pleasure of the Familiar speaks to. However, despite the human need for structure and order, familiarity is not the only source of pleasure. Being momentarily challenged, experiencing degrees of confusion, can be enjoyable and titillating. As a designer, it is essential to understand one's recipient or segment in order to understand how far to push the boundaries of one's design object or concept before the experience becomes unpleasurable.

The Pleasure of the Unfamiliar is often connected to an object's news value. It is far easier to provide people with a sense of excitement upon initial encounter than it is to maintain their interest—continually providing that particular pleasure of the challenging aesthetic experience. Nonetheless, it is by repeating or "expanding" on that initial pleasurable encounter that sustainability is produced. A major challenge of creating sustainable aesthetic expressions has to do with establishing *something more* than a brief bond between subject and object, a bond that is based on the immediate interest of news value. Aesthetically sustainable objects (or concepts) are things that people will derive pleasure from for years to come, and which, for this reason, they will want to keep and care for.

The next chapter deals with the components that together define the sustainable, or durable, aesthetic expression.

## Notes

1 See the section on "The beautiful" in Chapter 1.
2 Catharsis means "purification" in Greek.
3 Cf. the discussion about Johannes Itten's seven color contrasts in the section on "The universal effect of color" in Chapter 1.
4 Cf. Kant's three stages of the sublime aesthetic experience in Chapter 2.
5 See Chapter 2, Figure 1.
6 I will return to the concept of aesthetic value in Chapter 6; briefly, however, aesthetic value can be described as an expressive and sensuous form of sustainability. An object (or concept) becomes aesthetically valuable when it manages to provide the user/viewer with aesthetic nourishment.
7 Cf. my description in the section called "Homeliness" in Chapter 1 of walking barefoot through my childhood home, sensing the changing texture of the surfaces under my feet.

# 3    The expression of flexible aesthetics

Which expressions one finds appealing is largely influenced by the trends and tendencies of the current moment. Despite this, it may yet be possible to construct a set of guidelines for creating generally appealing or beautiful objects, which, additionally, have the capacity to provide the user with pleasure for years to come—thereby defying the connotative frames, basic assumptions, and culturally anchored preferences. The concept of aesthetic sustainability can help clarify how such guidelines can be defined and implemented.

The notion of flexibility is pertinent to consider in relation to aesthetic sustainability. Like objects marked by the Pleasure of the Unfamiliar, flexible objects maintain user interest either by prompting interaction or by providing a changeable expression. Moreover, the flexible aesthetic expression contains a potential ambiguity that might prolong the time of decoding and detecting, which would then demand of recipients that they invest a certain amount of time in experiencing it. The culmination of the aesthetic experience, or pleasure, in other words, takes time to develop.

But what is a flexible object? And what does it mean to create a flexible aesthetic expression? In the following sections, I will clarify the concept of aesthetic flexibility. In doing so, I will describe and analyze flexible usage, flexible materials, and flexible function as well as renewal and decay. In addition, the concept of "constancy" will be explored—constancy as the lasting, the unchanging, and the remaining.

First, however, I will return to the nature of beauty.

## Fleeting beauty

What is beauty? And what does a beautiful object look like? Being able to answer these questions unequivocally would be of great value to designers who aspire to create objects of mainstream appeal. But unequivocal answers are difficult, if not impossible, to reach—not least of all when dealing with something as complex as beauty. Beauty is a fleeting concept, which is either considered as something that fades with time or as something so closely connected to individual taste and subjective associations that it is impossible to say anything definitive about it. However, as the two previous chapters have shown, my hypothesis is that, despite individual preferences founded in cultural or temporal contexts, it is possible to set out general criteria or guidelines for the kind of elements that make up the beautiful object. Put differently, objects or expressions perceived as beautiful are to some extent universal.

According to the French poet Charles Baudelaire (1821–1867), whose poems concerned and were inspired by the growing cities of the late 19th century, urban life teems with poetic and adventurous motives and with a kind of beauty, which he terms *modern* beauty, and which is characterized by transience, fragmentation, asymmetry, and chance. Taking urbanity as his starting point, Baudelaire distinguishes modern beauty from classical beauty, which is characterized by symmetry and harmonious proportions (Baudelaire 1998).[1]

The modern artist wishing to "capture" and enunciate, or perhaps visualize, modern beauty must move around the city as an anonymous observer in the guise of the "man in the crowd," like the *flâneur* or the idler, to take in the constantly changing expression and look of the city—and, by extension, fleeting, modern beauty. Baudelaire's *flâneur* observes the world while existing in the midst of it; at the same time, he is anonymous and remains hidden from the world, and this is his source of freedom. The artist-*flâneur*, according to Baudelaire, should only write or paint in concord with what he sees and feels, adding a sort of phenomenological element to the process: sensing, not analyzing or concluding, should be the starting point of the creative process.

Nevertheless, in sketching the random and fleeting elements of the modern city, the true artist seeks something universal, something imperishable—the enduring, invariable part of beauty or the modern *in itself*, the very core of modernity: "Every old master has had his own modernity; the great majority of fine portraits that have come down to us from former generations are clothed in the costume of their own period" (Baudelaire 1964: 13). In other words, the true artist aims to penetrate the essence of beauty of the particular time period through experiencing fleeting or fragmented sense impressions (attempting to put these into words or form in some way). This can be done through the imagination, which Baudelaire sometimes referred to as the "queen of faculties" (see Habib 2005: 495).

For Baudelaire, the beautiful thus consists of, on the one hand, an "eternal, invariable element," and, on the other hand, a "relative, circumstantial element," which concerns the fashions of the day (Baudelaire 1964: 3). Without the enduring element, a work of art appears meaningless. But transience (and news value) is also necessary for modern beauty. A work has to contain eternal elements, but, concomitantly, it must be marked by the fleeting variability of the world. For this reason, the artist, in order to *touch* the viewer's soul, must seek to express eternity and transience, immutability and fragmentation.

This point about the duality of the modern art work can profitably be transferred to the design object. The object that a viewer or user deems appealing, interesting, beautiful, and attractive might very well, in the spirit of Baudelaire, contain enduring as well as fleeting elements. Or, it might consist of an element that is independent and unaffected by time and place—remaining interesting and relevant for a number of years—*and* an element that is attractive by virtue of its transience and its news value.

In the section about aesthetic decay later in this chapter, I will elaborate on the attractive object's combination of immutability and transience further. In the next section, however, I will dwell a bit longer with the nature of beauty to establish further the connection to affective bonds between subject and object.

### *The nature of beauty*

An object's aesthetic, expression, and sensuous qualities are decisive in determining the extent to which it is perceived as being attractive—and, in part, as being useful— by the subject. The Pleasure of the Familiar, and the kind of beauty associated with it, largely depends on the object's structure as useful, immediately decodable or detectable. In this manner, the immediately comfort-inducing, the useful, and the functional are tied together.[2]

An object's beauty concerns more than making the object appear immediately useful or appealing; the expressive part of the object can act as the primus motor in establishing an emotional bond between a thing and a recipient, or between an object and a subject:

> The aesthetics of a product can be very powerful because they are a key factor in creating an emotional tie with the object. Aesthetics can help transform a product from an uninteresting and unusable collection of functional components into a useful and attractive object that provides a meaningful benefit to people's lives.
>
> (Walker 2007: 142)

As this quotation from Stuart Walker's *Sustainable by Design* makes clear, aesthetics contribute greatly to the emotional attachments people form to objects or products.

Human beings form emotional attachments to certain things—perhaps because they remind us of something or someone, or perhaps because their appearance and sensuous qualities appeal to us and "fill us" in a way that other more "insignificant" things do not. Some things fascinate us in specific ways, and there are indications that this type of object-based allure, as we might call it, has to do with the aesthetics of the things in question. We are compelled to touch, wear, use, and own alluring things. Such alluring things possess a magical pull that is hard to explain because it is not rational, but rather related to feelings and sensations. It might be suggested, however, that the things that attract us substantiate our identities in particular ways; they somehow mark who we are, becoming attributes of our individual characteristics and preferences in the process. Such things are existentially related to things of sentimental value, as an emotional bond exists between the thing and the owner that ensures the preservation of the thing. We are less likely to dispose of objects that we are emotionally invested in; similarly, we are more likely to repair such objects than we are "insignificant" things that can more easily, and perhaps advantageously, be replaced by different, newer, better, or more interesting and appealing things.

At the same, however, there is a difference between things of sentimental value and things that establish an emotional bond to the owner through their aesthetic qualities or their beauty—and thereby their allure. One particular characteristic separates aesthetico-emotionally sustainable objects from things that are predominantly marked by sentimental value: things that establish a bond to the owner or user aesthetically may very well be new items. They may also be used, vintage objects or heirlooms, for instance. Because of the inherent narrative power of used objects—often expressed through signs of wear and tear—they can easily establish a special bond to the recipient or user. *But*—aesthetico-emotional objects, it is important to emphasize, may well be new things. In relation to this chapter's ambition of establishing strategic guidelines for how to create design objects or concepts, the sustainability of which comes from their aesthetic value, this is an especially compelling perspective. The central question becomes, then, which components or properties a new thing must possess in order for it to obtain aesthetic value and for it to consequently establish a bond with the recipient.

For a person to want to buy and to use (and not least of all, to keep) a thing, it needs to appear useful. But for it to attract interest, in an immediate as well as in a continuous way, the thing must also be attractive or beautiful. The value of the latter is often underestimated. People need beauty. We thrive in beautiful, harmonious surroundings and tend to buy, clothe, and surround ourselves with beautiful things. It appears that this has always been the case. Numerous examples exist of ancient excavated jewelry, decorated everyday objects, and wall paintings. We are *nourished* aesthetically when surrounded and captivated by beautiful things. This need for aesthetic nourishment is relevant when exploring which elements the aesthetically sustainable object consists of.

In our culture, beauty is often considered a fleeting thing—something that disappears as a result of decay, wear, or aging; something that is "crisp" and new. In this way, beauty is connected to the polished, to the not-worn, to the "pretty," and to the pure, aromatic, and proper. However, as will become clear in the following sections, there exist philosophico-aesthetic traditions that associate beauty with the uneven, unpolished, and corporeally challenging rather than with newness, and even with aging, decay, and wear. This concerns the aesthetics of decay—a concept favored by photographers who are attracted to ruins and abandoned urban environments—but also has provenance in relation to aesthetic sustainability. The aesthetics of decay celebrate decline! Aestheticians of decay like to show the beauty of peeling paint; tattered wallpaper; overgrown walls covered in wild, uncontrollable vegetation; and pockmarked rooftops infested with pigeon's nests and spider webs—or, in other words, the beauty that appears when letting something "be," allowing the ravages of time, wind, and weather to stamp objects and buildings.

Applying the aesthetics of decay—the love of torn, raw, and pitted things—to fashion design will invariably challenge what it means to be well dressed. This could be done by celebrating the patina of used textiles, or perhaps by emphasizing repairs, patches, and random combinations of pre-existing materials. In relation to interior and furniture design, the influence of a decay aesthetic might mean that peeling wallpaper and paint, pitted furniture, and bleached textiles would become favored materials and objects—or the most appealing and thus beautiful.

This way of thinking offers a distinctively different perspective on the nature of beauty. If beauty is fleeting and transient because it is associated with what is new and polished, which is often the case in our culture and part of the world, it is not sustainable. As Stuart Walker puts t in *Sustainable by Design*:

> This line of thought suggests that there is a fundamental and irreconcilable disparity between our conventional notions of beauty – what might be termed "outer" or extrinsic beauty, and sustainability. The word sustainability evokes ideas of longevity, continuity and endurance. "Outer beauty", on the other hand, is perishable and transient. It exists only for a short period, but fades with time.
>
> (Walker 2007: 58)

The sustainable object (or concept) endures; it is continuous; it consists of "something" that keeps it being useful and, not least of all, appealing and attractive. "Outer beauty," on the contrary, to use Walker's term, is fleeting and ephemeral. As it "fades with time," beauty becomes equal to newness. As a consequence, the news value of an object becomes more important than its aesthetic value, something that the fashion industry and conventional *fast fashion* has helped to encourage.

But can *durable* objects thus not be beautiful?

Certainly, they can be. It is simply a matter of definition. In a Baudelairian manner, the beautiful, or the durable, can be linked to the essence of contemporary beauty, which means delving into and analyzing the Zeitgeist. This kind of Zeitgeist analysis is an effective way of working with trend-analysis—and an effective way of establishing guidelines for creating expressions, material or immaterial, that "speak" to one's chosen recipient.

## *Zeitgeist analysis*

As the basis for conducting a Zeitgeist analysis to help guide the design process, Schein's culture model can be employed.[3] Schein's theory can help us understand the general assumptions or myths that characterize the Zeitgeist in relation to a design product. The insight produced hereby can help designers understand the sources of status and what is considered the "good life" according to the Zeitgeist and different target audiences.

As stated in Chapter 1, the section on "The easily decodable," general assumptions have much in common with what Roland Barthes refers to as myths. The general or basic underlying assumptions, in Schein's view, correspond to *Truisms* (Schein 2004); they are "truths," or taken-for-granted assumptions, that a particular group of people does not question. Such assumptions might concern work, child rearing, leisure, housing, family, or love—all of which are topics that, taken together, constitute the foundations of the good life. That children should have a say in family matters, or that romantic partnerships should be monogamous, or that work life should be governed by enthusiasm and blurred distinction between work and private life—these are examples of general assumptions. These kinds of assumptions are "truisms" insofar as they are difficult to justify or explain (as is it just the way things are!). For those involved, these are considered as a matter of course. Of course one should be enthusiastic about one's work; of course loving relationships should be monogamous; of course children should have a say in family matters. Anything else would seem unthinkable. Encountering people with opposite beliefs would then come to seem entirely strange or foreign, and might result in cognitive dissonance.

Characteristic of all general assumptions is the fact that they are ineffable, or at least highly difficult to put into words, which means that, when prompted, members of a given cultural group will struggle to define them. In other words, it will become apparent that one has reached the "core" of a group's assumptions when members are finding it difficult to argue for the truth of them. At best, a likely response will be: "That's just the way things are." Basic underlying assumptions, so to speak, are found where words end. They are difficult, if not impossible, to argue for. They are so integral to the individual human being and her relationships of belonging and identification that changing them would require a complete paradigm shift.

Societal trends, similarly, are based on different general assumptions. For example, the trend to bake one's own yeast bread or to wear homespun Icelandic sweaters might be caused by the myth or general assumption that time is precious. If time is considered a precious commodity, as something to be safeguarded, it is not strange to think that the time and energy it takes to bake bread or to knit sweaters—or perhaps just to buy expensive sweaters made by an elderly Icelandic woman, who spent 80 hours on it—becomes a source of status, which translates into trendiness. This kind of insight is valuable to the design process, not least of all when it comes to deciding on the type of aesthetic value to build into one's object or product. Chapter 6, "The value of aesthetic sustainability," will explore this in more detail.

Schein's primary focus of interest is organizational culture, and his analytical model is based on this. Nonetheless, his model is highly useful in analyzing different societal trends as well. According to Schein's model, artifacts are to be understood as the visible or sensorial level of culture, which means that artifacts register as the most obvious markers of any cultural group. Artifacts[4] correspond to all those objects that the subjects of one's analysis prefer to surround themselves with. These can range from clothing to decorations and everyday objects.

When using Schein's model as a blueprint for one's Zeitgeist analysis, things that are considered attractive, prestigious, and beautiful by the group in question at a given point in time should be registered on the level of artifacts. For example, the members of a group, simplistically put, might choose to buy vintage and craft clothes for themselves and their children, and they might care about "slow" objects and offline culture, immersing themselves in books, record players, and clay pots, but at the same time, they might also consider the latest gadgets and electronic devices as absolute prestigious must-haves (despite the fact that these are indubitably "fast" objects belonging to online culture). On this level of analysis, then, the task is to collect—preferably in the most literal, visual, and tactile sense—material that can be used to illustrate what members of the group choose to surround themselves with: what they consider beautiful.

The next level of Schein's cultural analysis concerns conspicuous values or espoused beliefs. These are values that form people's preferences—as, for instance, the preference for using ceramic bread bowls, record players, homespun sweaters, and streamlined gadgets. This analytical level, in other words, is interested in the kind of explanations group members are likely to offer in terms of justifying their preferences, as in the ostensible contradiction of surrounding oneself with both handmade artifacts and mass-produced devices. The answers people give will likely contain a mix of truisms and useful explanations to help form an understanding of their specific preferences: the sensuous beauty of a handmade, rather uneven ceramic bowl, and the crackling sound of a record player; or the vast social interface of modern gadgets and the slick, streamlined minimalism characteristic of smartphones, which also contain a sort of sensuous pleasure.

However, the one sentence that might synthesize and connect the level of artifacts and the level of values can only be reached in one way: by studying, observing, and analyzing all the material gathered as part of one's investigation. Only through synthesizing analysis is it possible to reach the basic underlying assumption that will explain and shed light on why members of one's target group behave the way they do and why they surround themselves with certain objects. This part of the analysis should aim to synthesize the core of what is considered good and beautiful by the group of people one's product is designed to reach. The result, or the synthesis, in regard to the above example might be: *beauty is found in the feelings and stories a certain object (whether handmade or mass-produced) encompasses*. This type of sentence can then act as a catalyst for creating a collage or a mood-board, which, in turn, can help trigger a design process, leading to a product or concept development. As a starting point, the analytical process outlined above is apt to lead to a design that will "speak" to the unconscious needs that guide one's target group.

A three-step analysis consisting of 1) the level of artifacts, 2) the level of values, and 3) basic underlying assumptions can thus trigger a design process that gets at the essence of the group's needs as well as the Zeitgeist. Additionally, this kind of analysis (see the above example) can help explain why people, as well as trends of late-modern society, can be difficult to categorize in an immediate way, in how they might "buck" against understanding. Indeed, why is it not contradictory at once to prefer "slow" objects, such as record players and clay pots, and to also stay totally up-to-date with the latest electronic gadgets and smartphone apps?

By employing a Schein-inspired analytical model like the one briefly sketched here, it becomes possible to define a way of segmenting values based on an understanding of the Zeitgeist, rather than based on traditional segmentation tools and trend analyses, which often focus on demographics, lifestyle, and consumption patterns. Relying on a Zeitgeist analysis makes it easier to detect fundamental connections between apparently quite different societal trends or between contradictory needs in the user or segment in question. Additionally, and this is especially important in relation to aesthetic sustainability, a Schein-inspired model can lead to insights regarding the nature of beauty or what beauty means to a certain group of a people at a certain point in time.

In this chapter, I have hypothesized that beauty consists equally of transient and lasting elements. Beauty consists of elements that are independent of and indifferent to time and place—and which, for this reason, remain interesting and relevant for years to come—*as well as* elements that are attractive because of their fleeting and new qualities. The point about using the Schein model is to gain insight, partly, into the lasting core and, partly, into the fleeting element of a given "beautiful" object. Equally, the aim is to discover ways of incorporating durability and transience into a particular object or concept.

Concerning the above example, on the one hand, the need for consistency and permanence can be meet through slow objects, which, in some way, constitute a time capsule, as they have looked and worked the same for years. On the other hand, the need for efficiency, transience, and novelty can be satisfied through streamlined, effective gadgets. The particular group of people mentioned precisely needs one and the other part of the aesthetic experience—slowness as well as expediency, or constancy as well as transience.

Designers can work on accommodating the bifurcated need for constancy and variation of contemporary consumers. One might strive to satisfy each need equally in one and the same object, concept, or collection—or one might choose to mainly focus on one or the other. In the latter case, one should bear in mind that users or consumers will need to experience the entire "palette" of a design, even while one's object or concept prioritizes one element of the beautiful to the detriment of the other, for whatever reason. I will return to this issue in Chapter 7, "Aesthetic strategy".

In equal measure, beauty consists of permanence and durability, as well as renewal and variation. Only through some kind of interaction between each part can an object stay interesting and attractive over the course of many years; it is this interaction that is at the seat of sustainability.

In the coming sections, I will look more closely at the concepts of permanence and renewal, as well as their reciprocity. Furthermore, I will define the category of *slow aesthetics*.

## Aesthetic decay, slow aesthetics

> We cherish permanence and find a calming pleasure in things being the way they always are, so as not to delay our swift progress through the everyday. But if Aristotle is right about sensation having its own kind of pleasure, there must exist another type of pleasure, namely one of variation. Furthermore, Aristotle claims that "some things please us by being new, but please us less later on for the same reason."
>
> (Thyssen 2005: 31, transl.)

This quotation from the book *Aesthetic Experience* [*Æstetisk erfaring*] contains a number of interesting points that I will begin this section by emphasizing:

1   The importance of permanence or recognition, which permit our daily dealings, and which Chapter 1 dealt with in detail.

2   The importance of variation, which makes us stop and notice our surroundings and the objects we find here, and which is related to the Pleasure of the Unfamiliar, the subject of Chapter 2.

3   The importance of combining permanence and variation, or renewal, as permanence in itself can lead to indifference and stagnation as well as a kind of hypnotic state, wherein we do not pay attention to the objects or our surroundings, as they have become *too* familiar. In the same breath, variation for its own sake, or in the name of renewal—which aims solely to please the recipient's need for the spectacle of newness—might also lead to indifference.

Fashion is by definition focused on new and forward-looking trends. But despite its forward-looking and new-is-good doctrine, fashion constantly borrows from the stylistic expressions of earlier times. It is often described as moving like a pendulum between two dialectical poles—as between, for example, the minimalistic, "purified" expression and the decorative, decadent, and "maximalistic"—a movement that, rather than being either decidedly forward-looking or exactly retrospective, should be considered stagnant.

Every season is defined by new colors, styles, and themes, whether in relation to clothes, interior design, or food. These are often (especially regarding clothes fashion) "dictated" by trend agencies, which report on the focus points of the coming season, based on an analysis of societal trends. For example, a theme could be "urban boheme" complete with flowing robes, flowery patterns in warm nuances, dark velvet suits, and broad-brimmed felt hats, on the basis of the idea that we are living in a time when the individual is striving for a free, relaxed urban existence, reminiscent of artistic 19th century Paris. This is grist to the mill of our throwaway culture, but it is not in keeping with aesthetic sustainability. If themes change every season, who would want to parade around looking like a mix between a 70s hippie and a 19th century dandy, when the current look is minimalistic, "tight," and influenced by black-and-white geometrical patterns?

So, what is the best way to circumvent this mechanism? Currently, there exist many interesting anti-trend movements that "dictate" slow design, slow fashion, or slow clothing—all of which challenge the whirling forward momentum of fashion and the constant impetus to consume. In fact, it has become trendy to preface everything with "slow": slow food, slow travel, slow living, slow parenting, slow shopping, etc. In a way, this is a somewhat self-contradictory move, as the intention is rather to think beyond the trend machine. Nevertheless, the slow movement, in relation to aesthetic sustainability, contains a number of important points. In point of fact, aesthetic sustainability can be considered part of the slow movement—the part, which could also be called *slow aesthetics*, that aims to put a brake on aesthetic pleasure (to keep it from dissipating, which would lead to casting aside the object that caused it in the first place and moving on to the next in line), or, rather, to make the aesthetically pleasurable experience *last* longer.

The slow movement, in brief, is about celebrating slowness and challenging consumerism. The focus is on "little, but good" —the idea that slowness results in an increased focus on one's surroundings and the objects found here, and that this

focus is a source of intimacy; physical, sensuous intimacy, that is, a kind of intimacy that might disappear when living life in the fast lane, to use an obvious cliché.
I consider the slow movement and the intentions behind it as an expression of a
paradigm shift, rather than as a passing trend like many others.

Aesthetic pleasure is not just about renewal, but also about repetition, as Ole
Thyssen points out (2005: 32). A worthwhile piece, or a good design object, can
provide the individual with either the Pleasure of the Familiar or the Pleasure of
the Unfamiliar, time and again. This is exactly what makes it slow—and valuable.
Establishing a "relationship" with an object—bonding with an object in the sense
that merely a glance or a touch is enough to elicit an all-encompassing feeling of
comfort or an inspiring feeling of seeing everything in a new light—means that
one will want to keep it around. Getting rid of it, finally, would signal that one is
"done" with it, that one has perhaps moved on to a different stage of life, wanting
to make room for new moods and new impulses—which is, of course, only natural.

The slow aesthetic experience is a continual source of aesthetic nourishment
and enrichment; an experience that is characterized by being both constant
(something one will want to relive) and regenerative, as it is a continual source of
new *nourishment*—whether it is the kind of nourishment that leads to immediate
enjoyment or to trembling pleasure.

To create sustainable objects that are aesthetically durable, and which thereby
possess an aesthetic slowness, one must thus *juggle* permanence as well as variation,
fixity as well as vitality, repetition as well as renewal. Only in combination is it truly
possible to *nourish* the experiencing subject *aesthetically* and to open up an aesthetic, potentially insightful, realization and for the individual to feel "at home" in
the world. The durable aesthetic object (or concept), in other words, consists not
solely of pure permanence, which might otherwise be an obvious conclusion, as
permanence and sustainability are closely related. The durable aesthetic expression
consists rather of a combination of permanence (and thereby the *enduring* or what is
temporally extended) and renewal, variation or energy (meaning that which is surprising and *uneven*, that which triggers the mind into movement and challenges the
senses). Permanence is expressed, for instance, by shapes and color combinations
that speak directly to the viewer, and that are easy to decode, or understand, and to
use; permanent objects can be quickly absorbed and put to use. This element is
essential in relation to aesthetic sustainability as it is of universal appeal,[5] which
partly concerns the joy of recognition and partly the pleasure of satisfying one's
human need for synthesis and structure. However, renewal and dynamic energy are
just as crucial. To keep the viewer engaged, the object must leave an impression so
great as to separate itself from the enormous amount of human-made things that clut-
ter our world. Achieving this means that the object must contain a certain degree of
renewal and dynamic energy. The dynamic part of the aesthetic sustainable object is
the part that makes the viewer take note, and that makes her come back to the object
(or concept), again and again, as she seeks the immediate pleasure experienced in
first encountering it.

Of course, not all objects should possess the same degree of variation, renewal, or
vitality. If the goal is to elicit the Pleasure of the Unfamiliar in the recipient, the

design object should be *charged* with surprises, irregularities, and dynamic asymmetry to a much larger extent than f the aim is to impel a kind of enjoyment that is in keeping with the Pleasure of the Familiar. But even objects belonging to the Pleasure of the Familiar should contain an element of renewal. If the aim is to create a design object valued for its ability to prompt comfortable feelings in an immediate way, it is not enough to simply copy an already existing object. On the contrary, it is necessary to temper the familiar with the regenerative, with variation. The regenerative element can be rather low-key, consisting, for example, of a material that adapts to the form without issue and without *inertia*,[6] and that thereby matches the form and the use of the object, but which is nevertheless somewhat different from the material usually employed to construct the particular object. It might even work better. Working on creating such objects can be described as an investigation of how much the core of an easily recognizable or easily decodable jacket, chair, lamp, or cup can "contain": how far can the expression or the form or the sensuous qualities of an object be stretched, while still evoking the Pleasure of the Familiar?

The aesthetically sustainable object (or concept)—and hence the object that manages to appeal to a human being who will watch it, use it, consider it, or touch it, *time and again*—has a core consisting of an easily recognizable or easily decodable element *and* of either some or a lot of dynamic movement, regeneration, and variation. This is to ensure that interest in the object will endure for an extended period of time—maybe even for decades—and that the experiencing subject will return to the object (or concept) in search of new aesthetic satisfaction, creating a bond to it in the process.

In order to "translate" the above to a set of strategic design guidelines, the first step for designers is to uncover what kind of aesthetic nourishment they wish to impart to the recipient: the immediately comfortable or the more challenging, boundary-breaking aesthetic experience? When this has been determined, in relation to considerations concerning recipients' habitual behavioral patterns and values, and in relation to their lifestyle and sensuous preferences and habits, it becomes possible to set out directions for the design process regarding permanence and renewal.

As the basis for a design process, the goal of which is to create an aesthetically sustainable product, it is thus useful to consider the following:

1   Is the purpose of the design, fundamentally, to please or to challenge the recipient?
2   Who is the recipient? What does her cultural "baggage" consist of, and how is it used to decode her surroundings (in a semiotic sense)? Or, what are her sensuous expectations of the world and its objects (phenomenologically speaking)?
3   How much regeneration can the *core* or the form of the design take? In the example of a chair, how far can its basic form be pushed without losing touch with one's recipient, given her "baggage" and expectations of what a chair is? At the same time, how is it possible to fulfill one's own aesthetic ambition to create for the recipient, for instance, the Pleasure of the Familiar? Or, to what extent is it possible to challenge or renew the core of "outdoor apparel" or "coat," and thus provide one's recipient with an aesthetic experience that can form the source for long-lasting aesthetic nourishment?

Renewal can be expressed by rethinking an object's function or material, or the object's idiom, with the purpose of improving the design experience and functionality. It can also be expressed in the composition, or in the *sampling*, of components that do not typically go together: a relatively soft rubber material is sampled with a core or form that is immediately recognizable as being a table, whereby the traditional hardness and stability of the table is challenged, at the same time as a flexible and malleable object without sharp edges is created, an object perfectly suited for a child's bedroom, for instance. This imagined object would be even more suitable if it had a surface that were easy to clean and a (perhaps organic) form that "spoke" to small children and their constant, sense-based exploration of their surroundings, time and again.

In other words, the aim here is to figure out ways of combining permanence and renewal to create products that appeal to one's target group, ensuring, in the process, that the immediate appeal of the products leads to a durable aesthetic experience, in the sense that recipients would want to keep coming back to it.

In relation to the Pleasure of the Unfamiliar, and the sublime aesthetic experience, the ideal balance between permanence and renewal is achieved by sketching and experimenting one's way through a series of investigations into how far the basic form of an object can be "twisted": how much can a chair's form, for example, be renewed to still ensure that users feel challenged and, at the same time, experience the *payoff* characteristic of the aesthetic experience?[7] The point is that even design objects that intentionally aim to challenge and surprise recipients, by using unexpected forms, combinations, or multifunctional elements, must contain the possibility of a "new order" or of a progression from chaos to a sense of order and harmony.[8] The potential of understanding, decoding, or detecting, and using must always be present then. Designers must consequently ask themselves how far they are willing to depart from the basic form of a sweater, for instance, while ensuring that the recipient will understand it as still being a sweater and be able to use it as such—or understand it as a different kind of clothing piece, depending on its multifunctional character (perhaps it can be worn as a scarf as well).

As described previously, the good design experience involves the potential of a payoff in the form of an understanding of the object, however challenging and "crooked" it appears. And this payoff necessitates an appropriation of the object, a process of making it one's own. As the consumer appropriates a product, making its expression part of her identity, *slowness* occurs. The slow, durable aesthetic amounts to a recurring source of aesthetic nourishment—the kind of nourishment that produces pleasure, time and again.

Regarding the Pleasure of the Familiar, which is inspired by the beautiful aesthetic experience when combining permanence and renewal, as a rule one should remain relatively loyal to the kind of idiom that will be familiar to the recipient. That is to say, instead of form experiments aimed at discovering how much a sweater's form can be "twisted" and renewed, the design process should rather look at how little to renew the sweater's form, if the goal is to create an object that will appeal to the recipient's sense expectations and that will thus continue to appear attractive and interesting for a long time.

Objects that produce the Pleasure of the Familiar are characterized by accommodating recipient expectations regarding idiom, materials, color combinations, and functionality. But the object should still contain an element, however small, of renewal, variation, or aesthetic stimulation in order not to appear inconsequential or too anonymous. Such an element could be in the form of intricately patterned lining, which appears out of place at first glance, but which is nevertheless pleasing to the eye, as it corresponds to Itten's scheme of color contrasts[9] and thus complements the color of the blouse. The aesthetic element of surprise could also be in the guise of appliqués of a slightly untraditional material, which nevertheless support both form and function.

The beautiful aesthetic experience, which fills the recipient with pleasure, is not an *anonymous* experience, but rather an experience of "coming home." This should be understood in the way that the encountered object meets, or rather exceeds, one's expectations: one would not have been able to imagine a more satisfying object experience. A shirt triggering the Pleasure of the Familiar is the kind of shirt that appears instantly comfortable and stylish at the same time. Or it could be the kind of blouse that supports one's identity by feeling "right" immediately, comfortable and aesthetically pleasing.

Designing an object that is capable of satisfying the recipient's aesthetic needs—both by challenging her basic assumptions or sensuous expectations and by doing exactly the opposite: embracing expectations—requires detailed knowledge about the recipient. If one does not know the basic cognitive and bodily assumptions of one's recipient, accommodating or, conversely, challenging them will be exceedingly difficult. If the designer has not understood the connotations one's recipient typically associates with flowery patterns, steel, dark wood, or furniture of round organic forms—as well as the bodily or sensual expectations that precede these connotations—it will be very difficult to either please or challenge her.

## When an object assumes its character

As part of her research project, Local Wisdom,[10] Kate Fletcher from the Centre for Sustainable Fashion at London College of Fashion has been examining sustainable fashion since 2009; or, rather, she has been examining the characteristics of the clothes that people tend to want to keep and care for (even repair and pass on) during an extended period of time. This kind of project is interesting in relation to aesthetic sustainability and, more specifically, in regard to developing concrete guidelines for creating aesthetically tenable, sustainable clothing. By discovering traits that recur in the different pieces of clothing that people show a particular care for, it then becomes possible to incorporate these into new designs.

Additionally, at the heart of the Local Wisdom project is a wish to lower consumption. The sensuous and emotional bond between consumer and product—which arises from the tactile or storytelling qualities of a product, detailing its making—is a prerequisite for slow fashion or slow clothing to convince consumers to disregard the undulating pendulum of fashion trends.

The method used by Kate Fletcher focuses on collecting stories about how people maintain and care for their favorite pieces of clothing, and this is a method that could easily be transferred to the guidelines for making objects other than clothes. Indeed, the Local Wisdom project can be used as a starting point for conducting design-anthropological surveys of people's relationships to everything from kitchen utensils and furniture to bicycles and toys. This kind of anthropological method could then form a solid foundation for working with durable, sustainable products (and even company concepts or the like).

The Local Wisdom website contains a plethora of inspiring points about the relationship between consumer and clothing:

> With our garments, as with our bodies, the passing of time leaves its mark. Our relationship with these imprints is complex in both domains. With clothes, we sometimes discard pieces because they are ageing, dated, jaded or worn; at other times we buy vintage or pre-distressed pieces, coveting that which looks old. Yet these both overlook the power and pleasure of marking the passing of time as it is recorded in our clothes; the forging of memories, building of knowledge, evolution of appearance.[11]

The essence of this quotation is that our clothes, as well as our bodies, change over time, but that this adds value to the clothes, rather than subtracts value from them. Furthermore, the important thing is that, at best, in relation to the objects we anguish over getting rid of, we are dealing with a form of *aesthetic decay*.

In Chapter 4, "Designing the temporal object," I will return to how the traces of time can add value to an object. But regarding the combination of permanence and renewal, it is important to note that durable expressions and what I have termed *slow aesthetics* contain a measure of *aesthetic decay*. This aesthetic decay increases the value of the object; it makes it more appealing and, yes, beautiful! Not beautiful in a brand-new, glittery, and trend-based sense, but rather in the sense that (if this has been built into the design), as it ages, it comes into its own, aesthetically. Aesthetic decay leaves traces and stories on the surface of the object in the form of scratches, stitches, patches, or irregular and worn spots that testify to the way in which the object has been used and thus to the preferences, habits, and identity of the owner. These "stories" are characteristic of the invaluable added value typical of aesthetic decay and slow aesthetics. This is a kind of decay that signals both permanence and renewal; or, put differently, aesthetic decay contains a well-formed core, which keeps the shape of the object, so to speak, and allows for the transformation of the object's "coming into its own" that results from use. This transformation is also what permits the owner of the object to make it her own, letting her identity come through in the object.

Decay *is* renewal—a perhaps contradictory sentence that nevertheless characterizes the aesthetically sustainable product, which ages gracefully and which possesses the germ of aesthetic decay as process. Decay equals renewal in the sense that aesthetic decay ensures the continued interest and fascination of the recipient. Decay spells renewal to the degree that decay changes an object over time and thus creates variation and dynamic range. Sowing the seed of aesthetic decay in an object, in other words, is a way to combine permanence and renewal; hence, it is a way in which to create an aesthetically sustainable object. Through aesthetic decay, an object can keep renewing itself, and, in this manner, it can potentially maintain user interest for years to come.

But what does aesthetic decay require of an object? How is it possible to design products that contain and even "bloom" with decay? One way to work with decay could be in relation to developing a fit with a flattering silhouette that in time changes with the body. Another way could be by creating furniture that keeps pace with the life of its owner, slowly adapting to her gradual personal development, aesthetically rather than functionally. Or a choice might be to create objects that *register* their use in the sense that they are, for instance, embellished by wear (and that this is part of its design, from form to the choice of materials). Incorporating into an object the potential for change does not necessarily concern multifunctionality; rather, the idea is to create objects that are capable of aging aesthetically, and that only become more beautiful with time. Similarly, objects that support our unique identities for years, or perhaps even for life, because of their gradual decay express this particular design attitude. Additionally, a strategy of producing aesthetic decay could lead to objects that are suitably "neutral," meaning that they would be able to enter into a range of contexts and thus appear beautiful in combination with other objects.

A different option could be to find inspiration in other cultures that have a tradition of repairing, maintaining, and caring for objects—a tradition which sees the wear and tear of time as being beautiful (and regenerative in the above sense). An example of such a tradition would be Japanese culture and the tradition of *wabi-sabi*.

## Wabi-sabi aesthetics

In *Wabi-sabi for Artists, Designers, Poets & Philosophers*, American architect and design theorist Leonard Koren (b. 1948) writes about his "encounter" with the wabi-sabi philosophy: "Wabi-sabi resolved my artistic dilemma about how to create beautiful things without getting caught up in the dispiriting materialism that usually surrounds such creative acts" (Koren 2008: 9). Wabi-sabi is a Japanese aesthetic philosophy celebrating the simple, restrained, perishable, and rustic; it is closely associated with Zen Buddhism. Wabi-sabi is often also connected to Japanese tea ceremony, where all actions are choreographed, concentrated, and slowly orchestrated, and where the objects used have been carefully selected based on an aesthetic ideal that prescribed rustic, restrained, and simple elements. Irregular, obviously handmade, cups and bowls are considered the most valuable, and hence the most beautiful, objects of the tea ceremony. The beautiful is what reminds of nature, and thus precipitates meditation and mindful presence. And since nature does not consist of smooth surfaces and polished perfection, but rather consists of organic shapes and rugged, tactile, stimulating planes—as well as an interaction between symmetry and asymmetry—these are the expressions idealized by the wabi-sabi aesthetic.

Objects created according to wab -sabi aesthetics and design principles should reflect human life and nature's perishable and transitory elements. Hence, it is important that such objects consist of "crooked" and uneven properties; they should in some way appear imperfect. Only imperfection contains true beauty, according to wabi-sabi. Additionally, wabi-sabi objects should ooze, or be *charged* with, slowness, so to speak, as well as invite use—an invitation that can be expressed by tactilely stimulating surfaces and natural materials, which become increasingly beautiful or progressively exquisite to touch over time and by being used (as do, for instance, natural materials like leather and wood):

> The physical decay or natural wear and tear of the materials used does not in the least detract from the visual appeal, rather it adds to it. It is in the changes of texture and color that provide the space for the imagination to enter and become more involved with the devolution of the piece. Whereas modern design often uses inorganic materials to defy the natural aging effects of time, wabi sabi embraces them and seeks to use this transformation as an integral part of the whole.
>
> (Juniper 2003: 106)

Wabi-sabi objects embrace the traces left by the wear and tear of time and common use; these are even considered beautifying. Wabi-sabi practitioners believe that time—even years—adds extra depth to an object, and such aesthetic preferences clearly enhance an object's durability. Chapter 4, "Designing the temporal object," deals with how the ravages of time and the use process can *charge* an object with stories in order to establish a strong emotional bond between subject and object. This notion is inspired considerably by the wabi-sabi philosophy.

The wabi-sabi object is asymmetrical or irregular and, to some extent, governed by randomness. The random expression can occur when the artist or the designer engages her materials in a "dialogue" during the process of creation; this is to be understood in the way that the artist or the designer should let the material "lead" the process, rather than focus on an overarching, conceptual idea. According to wabi-sabi, the designer should therefore detach from rational and structured thoughts during the experimental part of the design process, seeking instead to be carried away by her intuition and by the material between her hands. We find here a link to phenomenology. The thought of committing to the material and its "will" as well as one's own tactile sense coheres with how phenomenologists consider the sensuous, pre-rational interaction with worldly phenomena as being a primary source of insight. Furthermore, this way of thinking is similar to Ørskov's. In *Detecting Objects* [*Aflæsning af Objekter*], Ørskov points out that if a sculptor considers her job to be the creation of symbols, she will invariably proceed in a backwards fashion, with the risk of creating an object that can never be more than a fad, fleeting and unsustainable (Ørskov 1999: 101). The sculptor must enter into a "dialogue" with the material in order to create sustainable objects that invite to detection rather than decoding and interpretation.[12]

"Wabi-Sabi is a beauty of things imperfect, impermanent, and incomplete. It is a beauty of things modest and humble. It is a beauty of things unconventional," writes Koren (2008: 7). In addition to the above-mentioned elements, the imperfect, uneven, "crooked," unfinished, and asymmetrical are valuable aspects of wabi-sabi aesthetics to consider in relation to aesthetic sustainability.

Considering the unconventional as being valuable and beautiful reminds of the Pleasure of the Unfamiliar and the sublime aesthetic experience. For something to be unconventional means that it turns convention, the traditional, and the "usual" upside down, including our expectations concerning our world and its objects. The unconventional shakes our basic assumptions and our connotations.

The unconventional object can be a regenerative object that, in some different way—and staying true to the wabi-sabi approach, preferably in a simpler way than usual—offers human beings a solution to a practical problem. Or it can be an object that consists of untraditional combinations of materials or of a different surface finishing. As part of the impetus to turn expectations upside down and the preference for the unconventional, the wabi-sabi aesthetic prolongs the time it takes the recipient or the viewer to detect or to understand the object. This is related to the pleasure associated with the rush only afforded by transcendent and horizon-expanding experiences.

Durability and the unconventional are thus closely related. Objects created in accordance with the wabi-sabi aesthetic depart from conventional ideals of beauty, which are often associated with fleeting trends and, consequently, a low degree of sustainability. The aesthetic pleasure of wabi-sabi, then, must be considered as something different from "normal" or comfortable, but insignificant, experiences of beauty. In line with wabi-sabi philosophy, the true aesthetic and pleasurable experience should comprise a psychological movement or development of some kind.

Moreover, humility is a central element of wabi-sabi aesthetics. The focus is on details, chance aesthetics, and tactile variations to give the recipient a subtle, intimate sense experience that encompasses the raw, irregular elements of nature. Colors should be muted and based on natural dyeing. Chance, or randomness, should govern the process, as great conceptual ideas and an abundance of forced meaning and symbols are not highly regarded. Rather than force meaning into an object, the artist or the designer should let the material "speak" and guide her: "Quite often what is not added is more important than what is" (Juniper 2003: 107). The combination of humility and the unconventional are closely related to my own thinking that aesthetic sustainability is to be found in the combination of permanence and renewal.

Wabi-sabi aesthetics challenge our usual (western, contemporary) view of beauty. Additionally, the type of beauty prescribed by wabi-sabi is one of durability. Seeing beauty in the imperfect, the humble, and the unfinished runs counter to western consumerism. In western culture, beauty is often equated with the polished and the ostentatious—and the finished. We are not interested in unfinished products; process does not appeal to us, which is why it is rare for a western designer to disclose and share her creative process. An object or concept is not presented to the world until it is polished and ready.

However, a growing interest in the process or the time behind a product is taking shape. I consider this interest as part of the paradigm shift I mentioned earlier in this chapter in relation to the slow movement—a shift in the Zeitgeist that involves an increasing, joyous interest in products that have been handmade with care from the bottom up.

This paradigm shift is leading to a change in consumer culture, which, among other things, can be seen in the growing interest in sustainability in the form of durable products that are more or less timeless. What is more, a change in what counts as being beautiful is occurring. I will come back to this in Chapter 6, "The value of aesthetic sustainability."

## Sustainable thingness

> True measures of "durable" product lifetime are best found along emotional and cultural indices—what meaning the garment carries, how it is used, and the behavior, lifestyle, desires and personal values of the wearer. These empathetic connections are already well explored and understood by comp-anies, since they form the very basis for marketing strategies to sell more products. Using this information not only for financial gain, but also to direct design for emotional attachment to optimize product life for sustainability gains, is quite unfamiliar and uncomfortable territory. It challenges the very core of existing business models.
>
> (Fletcher and Grose 2012: 85)

As this quotation from Kate Fletcher and Lynda Grose's book, *Fashion & Sustainability* (2012), makes clear, seeking to prolong the life of objects or products (whether through qualitative, emotional, or aesthetic means) meets with certain challenges that require a new mindset.

Establishing an emotional bond between object and subject—between product and consumer—entails a design strategy devoted to nurturing, maintaining, and caring for things, rather than one aimed at seducing consumers by drawing on the fascination with "perfection," the new and the polished. To my mind, seduction is not an entirely negative term. Human beings want and love to be seduced; seduction contains a particular kind of devotion and satisfaction. However, the constant search for something new, something different, or something more exciting involves a degree of indifference and meaninglessness. Immediate fascination is often tempestuous, but just as often it is fleeting and this is not, in itself, durable:

> The passionate early stages of a subject–object relationship could be described as a honeymoon period, a period of intense synergy within which everything is new, interesting and the consumption of one another is feverish. Honeymoon periods are by their very nature short lived and must, ultimately, give way to the inevitable onset of normalcy.
>
> (Chapman 2011: 63)

What Jonathan Chapman here defines as "normalcy" can be used to describe our everyday relationships with things that, despite the ostensibly negative connotations, are built on respect and continual benefit; in fact, such relationships are similar to an affirming and loving relationship between two people. "Normalcy," which ought to be the natural "next step" following seduction or the "honeymoon period," only rarely occurs regarding most subject–object relations. Why? Because we have become accustomed to the "new", by definition, being the best, the most interesting, the most giving, and the most *beautiful*. And because, consequently, we consider buying new things or replacing old objects—deemed "over the hill," outdated and worn out—a pleasurable act. As Chapman goes on to say: "During recent years, consumers have become serial honeymooners, and today subject–object relationships are less marriage, more one-night stand" (Chapman 2011: 63).

Durable relations between subjects and objects, or between people and things, are suffused by normalcy, the everyday and the continual repetition of routines. In this case, repetition becomes the source of recurring joy or aesthetic pleasure.

Only in durable subject–object relations characterized by continuation, and hence normalcy, can the decay of things be experienced as being aesthetical or beautiful. Decay concerns the stories innate to any thing, and is thus crucial to its character and to its continued power of attraction. Based on Chapman's metaphor about the early stages of subject–object relations (falling in love and the honeymoon stage), the kind of attraction concerning normalcy or decay can be compared to a loving relationship between two people. In the same way that the scar on a lover's leg or the wrinkles around his eyes appear beautiful or characteristic of the person because they contain the stories about the time he fell off his bicycle and all the times he has laughed and smiled, aesthetic decay possesses qualities that make an object more valuable.

In the durable relation or bond between a subject and an object, between a person and a thing, their separation ceases to exist, in a sense; a pair of well-worn jeans, for instance, might come to seem part of the owner as a form of *second skin*. They become nearly a corporeal part of the person wearing them—exactly because of their wear and tear—which is why she will want to keep them for as long as possible and even resort to repairing them.

The aforementioned project, Local Wisdom, employs an investigative approach to combining ethnographic methods and the design process with the purpose of establishing how people from different cultures use, bond with, maintain, and repair their clothes. The project's website details stories about people from the U.S., Australia, Canada, and a number of European countries who have bonded with specific pieces of clothing, which they are loath to discard, and which they will repair if necessary because they love wearing them so much.

There are certain characteristics common to these special clothing pieces, as portrayed on the website. Among others, some are that:

1   They are worn often and have never been washed (even if they are made from materials other than leather).
2   They have been passed down or bought used.
3   They can be repaired easily.
4   They amaze on every occasion.

5   They are flexible and can be used in many different ways.
6   They are made or designed to meet differing needs.
7   They show and tell stories about how they have been worn.

These selected characteristics for durable clothing are all related to aesthetic sustainability: they support and supplement many of the points already made in relation to how to characterize an aesthetically sustainable object. Drawing on the above characteristics, the following section will thus present some strategies for creating durable and long-lasting products.

### *1. The conscientious, well-dressed consumer*

As mentioned previously, aesthetic sustainability refers in part to what it means to be well-dressed, and what *nice-looking* means. Inspired by the Local Wisdom project, an aim for sustainability designers could be to create products—in this case clothing items, but this line of thinking could be expanded to include other, different product categories—that never, or only rarely at least, need washing or cleaning. Doing so would mean including signs of wear and tear—and the kind of "dirt" that would naturally result from not subjecting the item to cleaning—into the aesthetic expression of the product. This could be one strategy of sowing the seed for establishing a strong emotional bond between subject and object (over time, personal use would create and change the expression of the object, which would thus become part of the subject, in a sense). In this way, the object would become *charged* with time, a concept I will return to in the next chapter.

What it means to be well-dressed is culturally specific and therefore something designers can influence. Hence, being well-dressed need not necessarily connote wearing "clean and freshly ironed" shirts, for example; it could as well come to mean being a conscientious consumer. In this way, being well-dressed could amount to wearing clothes that have been ethically produced, using environmentally sound processes. Such a connotative shift would require a specific and sustained design focus, of course, which concerns the communicative aspect of design to a greater degree than it does the product itself. The consumer must be informed and made interested in the process behind a given product. I will return to this in the section "Communicating sustainable aesthetic value" of Chapter 6.

### *2. Heirlooms or used things*

Second-hand or vintage objects, as well as heirlooms, have the specific quality of containing many innate stories and "traces" that can be difficult to imitate when designing new objects. Often, these traces of use establish a special bond connection with the recipient or user. Additionally, the type of aesthetic value associated with used objects has the effect of seeming unique, one-of-a-kind. The value of discovering a unique and beautiful bowl, a wonderful shawl in strong striking colors, or a beautifully shaped hat in a second-hand store or at a garage sale, or the value of inheriting a necklace, a watch, or a handmade sweater from one's grandparent is akin to making a great find, a treasure that is like none other. Making such a find, or coming into possession of a "treasure," one will undoubtedly never want to part with it. If one does, it will most likely be to pass it on to someone very special. Such an object must certainly be considered aesthetically sustainable.

### 3. Easy to repair

The notion that durable objects are things that are possible to repair is simple and concrete. If it is not possible to repair an object, it is of course not possible to keep using it should it fall into disrepair. For this reason, it is crucial that repairing it is an actual possibility. It is important for designers to consider this as early as the initial design process, if they want to create durable products. Designing the potential of repair requires more than supplying jackets with extra buttons, for example; designers must fundamentally consider how recipients can easily maintain and repair products, perhaps in a fun, or even joyous, way. This will be a welcome addition to many types of products, not least of all toys.

Another way of working with products that are easy to repair or maintain could be to make them less *polished* (see the previous section on slow aesthetics). Objects that appear highly polished will come to seem "tired" fairly quickly; further, they can be difficult to repair, as minor repairs—in the form of patching, stitching, or coloring (changing the original nuance slightly)— can easily make the object look blemished. However, if the original design draws on the wabi-sabi aesthetic, repairs will only add to the already irregular and varied surface structures, contributing to the overall sensuous expression.

### 4. The continual effect of surprise

Objects that appear surprising every time they are used—and which because of this innate potential to surprise must be considered durable, as recipients will want to return to them again and again—posses the "ability" to provide recipients with a particular kind of aesthetic pleasure, that is the Pleasure of the Unfamiliar. These might be objects that challenge universal aesthetic principles and which thereby challenge the individual senses, or they might be products that turn conventions on their head, or just slightly disturb recipients' basic assumptions.

In this regard, one might ask why, in our time and culture, we have become used to always expecting something new to happen; why the new, by definition, is always the most interesting, the most exciting, the most rewarding and, not least of all, the most beautiful? Is it not possible to be seduced by the same object or product, time and again? Or one might ask: what would it take to be seduced by—fall back in love, become fascinated all over again, with—a familiar object over the course of a decade?

### 5. Flexibility and multifunctionality

The idea that flexible objects, which can be used in different ways, are durable is something I have touched on previously. Nevertheless, I consider the most durably flexible that which combines renewal with permanence.[13]

In relation to the Lost Wisdom project, the focus is on ways of incorporating multifunctionality into the object and on how to provide consumers with the possibility of using the product in unique and individual ways, in order to *customize* it. Additionally, the idea is to create clothing pieces that are neutral, but still flattering, so as to make them flexible in the sense that they can be used in a number of different contexts and can be paired with a wealth of accessories to change their expression (dressing them up or down, so to speak). This notion that the most durable expression is neutral and restrained is similar to the kind of durability and aesthetic sustainability associated with products that elicit the Pleasure of the Familiar.

*Sampling* functions that traditionally are spread out over a number of different products would be yet another way of working to create multifunctionality. One could find inspiration in an obvious example of multifunctionality: our indispensable smartphones with their combination of phone with camera, notepad, GPS, and so on, thereby collecting into one product or concept a series of functions and services that would otherwise be divided between different products.

## 6. Accommodating changing needs

In relation to aesthetic sustainability, it is imperative to bear in mind the role of the designer. Products ought to be *designed* to accommodate the changing needs of users. The object that follows and develops with the user throughout her life must be considered extremely aesthetically sustainable. Such an object could be a piece of furniture that can be made to change form, that is self-cleaning, or that can easily be re-upholstered; it could be a piece of clothing, the style and form of which can be manipulated to accommodate the body as it changes and ages; finally, it could be an everyday object that can be used in different ways according to the nature of the need or the life-stage of the owner. In this regard, I cannot help but think of a small bowl decorated with tractors that I received as a gift when my son was born. This bowl, having first aided in serving mash to the little guy, has, in turn, held pearls, snacks, herbs, water for dogs, and soup, to name but a few. Its form is perfect: slightly wide and not too deep, and the tractors are appealingly retro-naivistic in expression. The bowl is useful and its playful expression *lasts*. And this despite the fact that, I am pretty sure, it was not designed with an eye to aesthetic sustainability.

Generally put, objects that have been designed to accommodate changing needs are things that endure in terms of touching, using, looking at, and combining with other things over the course of many years; they are objects one would want to move with and keep throughout life; they are an endless source of aesthetic pleasure and nourishment, whether in the guise of the Pleasure of the Familiar or the Pleasure of the Unfamiliar.

In the section on aesthetic decay and slow aesthetics, I introduced the idea that by sowing the seed of aesthetic decay in an object, it becomes possible to work with both permanence and renewal, thus creating an aesthetically sustainable object. The notion that, as part of the design process, it is possible to work into the product or object a potential for change is crucial in relation to those elements that make up the durable expression. However, there is a difference between working with aesthetic decay and working with products that will accommodate the changing needs of users. Whereas the gradual and continual satisfaction of the object's owner requires a thorough understanding of the target group—in terms of identity, values, and basic assumptions (regarding "truth" and the good life)—working with creating objects of aesthetic decay requires a more phenomenological understanding of the object and the subject. Creating a foundation for aesthetic decay demands an expert understanding of material and form (regardless of it being in relation to a chair, a jacket, or a cup) and the ability to enter into a near-symbiotic investigation of how precisely this form and this material can absorb and reflect wear and tear in the most beautiful, most interesting, and most sensuously satisfying way possible.

## 7. The innate story of use

The idea that clothing that shows and tells a story about how it was made is sustainable serves as a natural segue way to the next chapter, "Designing the temporal object." That chapter will present three ways of "implementing" time into a design in order to create more durable products. In addition to the story about how a thing has been used and the durability thus produced, the chapter will look at the sustainable aspect of the story about how an object was made or produced as well as the story about how it is "to be with" the object.

In the book *Fashion & Sustainability: Design for Change*, the source of this chapter's epigraph, Kate Fletcher and Lynda Grose describe a number of ways in which to work with flexible clothing, including "transfunctional, multifunctional, trans-seasonal, modular garments", as well as "garments that can change shape" (Fletcher and Grose 2012: 77–83). Garments that can be used for more than a season—in the summer and in the winter, for example, by incorporating removable lining or zippered sleeves to a jacket or by introducing modular components that can be replaced or combined by the consumer—extend the lifetime of a standard wardrobe. Flexible clothing minimizes the need to change and renew a wardrobe. As part of their analysis, Fletcher and Grose consider the fashion designer whose role will change if the current emphasis on trends and seasons is replaced by an ideal of durability.

As consumers to an increasingly greater extent want sustainable and flexible products, which, in terms of quality and aesthetics, can last for years to come, designers who are used to creating products that neither can nor are meant to last for long will need to reconsider their approach. For this very reason, in this book I focus on aesthetic value, strategies for planning the aesthetic design experience, as well as aesthetic sustainability.

## Notes

1  See also the description of the Pythagorean view of beauty in the section about "The beautiful" in Chapter 1.
2  Remember the idea that a beautiful chair is good at being what it is, namely a chair (see the section on "The beautiful" in Chapter 1.
3  See the section on "The easily decodable" in Chapter 1 where I introduce Schein and his theory.
4  The word "artifact" comes from the Latin *arte factum*, which means "what has been produced as the result of a craft or art"—i.e., objects, or processes, as a result of human activity.
5  See the section on "Adhering to universal aesthetic principles" in Chapter 1.
6  Cf. Ørskov's use of "inertia" in the section on "The experience of minimal inertia" in Chapter 1.
7  See the section on "The sublime" in Chapter 2.
8  See "The stages of the sublime" in Chapter 2, where I describe the *Bildung* of the aesthetic experience: order–chaos–(new and improved) order.
9  See "The universal effect of color" in Chapter 1.
10  www.localwisdom.info
11  www.localwisdom.info/gallery/view/360/back-mending
12  See "The experience of minimal inertia" in Chapter 1, where I mention further the difference between detecting and decoding an object.
13  See the previous section on "Aesthetic decay, slow Aesthetics."

# 4 Designing the temporal object

Designing objects with temporality in mind can be a way of creating an emotional bond between an object and the recipient or user; it can also be a way of creating the potential for durability and aesthetic sustainability. As an object becomes a *container* of time—and thus physical, material, or concrete stories—it is *charged* with emotional and tactile value, making it more than just a thing. In a sense, the object is transformed into a qualifying capsule that has the ability to precipitate a momentary travel in time, opening up corridors to lead the subject back to, or into, hidden or forgotten sensations, bodily memories and feelings.

The previous chapters have focused on the aesthetic *experience*. In this chapter, however, and the next, I will focus on the temporal design object and the magical thing, respectively, in order to highlight the value of the thing-in-itself. Focusing on temporality means that the object comes into relief, rather than the previously described interaction between subject and object. In this way, the object becomes more important in relation to the aesthetic experience than the subject's cultural "baggage" and personal frame of reference.

The recipient experience is difficult to control. It is possible to analyze which elements will "trigger" recipients, leading to insights regarding how materials, structures, and forms correspond to a sense of familiarity and comfort, for instance, for a given audience or segment. Further, culturally determined factors, such as trends and values, can be used to incorporate signal value into one's product to thereby "encapsulate" the tastes and preferences of the target group. But it is impossible to be completely sure that the product or object will be decoded in the intended way once it is released into the world.

By contrast, if substantial aesthetic value is placed on the object itself—thus placing less emphasis on the subjective experience and interpretation of the object—it becomes possible, to a greater extent, for designers to create something durable. Circumventing the difficult to control subject–object relation by focusing instead on adding a sense of time, or other qualities, to the object itself creates the potential for long-lasting and durable design. In other words, object-oriented design turns its back on the aesthetics of reception and the idea that meaning happens as a result of the interaction between work and recipient. Instead, the attention is shifted to what can be called "thing-magic", whereby I mean the kind of magic that inheres in the thing-in-itself.

As the sender or creator of an object or concept, it is impossible to ever entirely get beyond the fact that the cultural "baggage" of one's recipient will influence her understanding, decoding, or detection of the product. However, it is possible to minimize the impact of these factors (which are undeniably difficult to control) by striving to create objects, for instance, that contain immanent "stories" or that possess a form of universality, which might help eliminate the many different decoding and detecting possibilities. This kind of universality contains a measure of aesthetic sustainability—a durability beyond time and place.

A way of *charging* a design object with time could be to implement the time of becoming or to make visible the design process in the object in order to establish a "relationship" between object and subject. The raison d'être of this kind of relationship is the story about what came before the object's three-dimensional spatial existence. The story about the time of becoming can appear as an instance of verbal storytelling, focusing on the process, methods, or techniques that have been used in the design process (a story that could be reproduced on hangtags or on the designer's website), or it can be told by implementing *traces* of the process into the finished product through making obvious the manipulation of the material, or by integrating handmade elements—embroidery, lace, or knitted details, for example—into the product.

The latter method of charging objects with time is particularly fascinating, as it entails the object becoming a carrier of time, literally; or, that is to say, the thing-in-itself appears as being a time capsule.

Another way in which the object might become a container or carrier of time is as a result of its decay. As discussed in the previous chapter, decay can be aesthetic; it can add value to an object and thus a kind of beauty, characterized by being "imperfect," unpolished and random, due to the ravages of time. This "ravaging" will in some cases embellish the object, making it more interesting, fascinating, and attractive. Wear and tear leads to imperfections that might make an object more tactilely stimulating and, simply, more *beautiful*. The wear and tear of an object tells the story of its use and the time that has passed.

To an extent, aesthetic decay can be imitated or copied, making it into a formula; not fully, however, as the authenticity of the object, when imitated, will necessarily dissipate. Nevertheless, designers can make it so that created objects decay or wear in an aesthetically durable way, meaning in a flattering way. In this manner, objects can be charged with time in order to attain emotional and aesthetic durability.[1]

A third way of adding a temporal dimension to objects is closely related to the Pleasure of the Unfamiliar, and thus concerns the time it takes to detect, or apprehend, an object, which demands being physically present with the object. This form of temporalizing an object might seem more abstract than the two other methods described above, since we are not here dealing with either the time it took to create an object or the amount of time it has existed, but rather with the time it takes to "take in" an object. How much time passes, in other words, from when a subject "meets" an object until she has apprehended, or taken in, its structures, shapes, surfaces, density, etc.? Crucially, this way of temporalizing objects has to do with the subject's phenomenological interaction with things.

In *Detecting Objects* (1999), Willy Ørskov (1920–1990) deals with three concepts of time that are highly relevant in regard to exploring aesthetic sustainability: the time of becoming, the time of existence, and the time of being. These three concepts have served to structure the discussion about how to work with charging an object with time. In my interpretation of Ørskov's concepts of time, they can be used as guiding principles in relation to the design process if, as a designer, the wish is to *charge* an object with time, thus seeking to create a durable connection between subject and object. Charging a design with time can be one way of working with aesthetic sustainability, then. Furthermore, the concepts of time can be used to analyze and apprehend, or decode and detect, design objects and concepts.

Ørskov's phenomenological point of view means that he considers physical objects, and what shows itself to consciousness (*phenomena*), as the most crucial path to insight. The body and the senses, in other words, constitute the most vital point of entry for human beings to navigate and understand the world. And the object (which for Ørskov concerns sculpture primarily, but which in my interpretation can also encompass the design object more broadly conceived) is a source for understanding more about life and the world—through the body and by way of the senses, that is, rather than through cognition or reflection.

The aesthetic experience based on phenomenological insight is neither intellectual nor founded on thought in any way; instead, it is corporeal and sensuous. In this regard, it is worth pointing out that the Greek word *aisthetikos*, the origin of "aesthetics," really means "to sense." In discussing aesthetic experience, one is really dealing with sensuous experience, according to the original meaning of the word, at least.

The symbolic side of an object—meaning the connotations and associations that an object might trigger—for Ørskov is only a secondary quality, whereas the primary quality concerns the purely physical and spatial existence or presence of an object. It is thus by *being-present-with* an object that the world becomes available to the subject in a new way, rather than by interpreting its symbolic value.

If a sculptor (or a designer), as the point of departure for the creative process, has decided the "fate" of the object based on its symbolic value, she is then, according to Ørskov, working in a backwards or reverse manner, and is thereby at risk of creating fleeting objects that can only be described as what Ørskov calls "fashion phenomena" (which is a strictly negative designation!). The durable and truly effective object is in conflict with fashion phenomena, or what could also be called trends.

The durable, aesthetically sustainable object can provide users with a sensuously insightful experience. As part of this kind of experience, the recipient is confronted with her bodily expectations of the world—in turn, these will either be confirmed or repudiated. Once more, we have returned to this book's two main categories of the aesthetic experience: the Pleasure of the Familiar and the Pleasure of the Unfamiliar.

In the following sections, my interpretation of Ørskov's concepts of time will act as the analytical turning point; more broadly, Ørskov's temporal insights will form the basis for the ongoing discussion of aesthetic sustainability in the remainder of this book.

## The time of becoming

According to Ørskov, the time of becoming is basically the time it takes for a given material to assume the form of the finished object. He describes this temporal process in the following manner: "The time of becoming concerns the degree of force and velocity as expressed in the encounter with the substance: the progression of creation, or the rhythm of composition—growth" (Ørskov 1999: 85; transl.). The longer it takes to create, the more complex the final expression, *typically* (even though this is not always the case; at times, the flow of a moment can be realized in the shape of a complex composition or illusory form; as well, the slow processing of a material can result in a simple or easily accessible expression). In this context, the time of becoming has to do either with implementing the creation process in the object or with telling the story about the time and the process behind the making of a product.

Involving the recipient in the process of making a product can help create a sustainable bond between object and subject, as the story about the product's creation, on the one hand, "speaks to" the preference for slowness, characteristic of our current moment,[2] including things and concepts that have been carefully crafted and that exude a surplus of energy and thoroughness; the value of a product is increased when the time of becoming is prolonged because of a comprehensive processing of materials and/or a series of careful considerations regarding the development of the product. On the other hand, in a more universal and timeless manner, the recipient can be involved in the processing of materials and in considerations regarding form by stimulating her senses to engage in tactile and visual investigations of the object, hence increasing her degree of involvement in the object. Designers can charge their object or product with the time of becoming, with process. In this way, it becomes possible to implement the rhythm of composition, by which I mean the inclusion of the process the material has undergone to assume the intended form (or the form that the process has led to) in order to make it stand out clearly and be easily accessible.

"An object is a frozen or fixed event" writes Ørskov (1999: 77; transl.). It is exactly the event or the actions behind the object that one's recipient is included in by charging it with the time of becoming. In this way, the object changes from being a static three-dimensional thing to being an experience and story, which appears a certain way the moment it is caught by the recipient's gaze and hands; the appearance of the object is conditioned by its *past*, but, like an organism, the object also has a *future*, meaning that it will continue to exist and evolve.

Previously, I have discussed the aesthetics of decay, and, in that regard, I introduced the idea that decay contains a degree of renewal. This is to say that an object that has been created to age in an aesthetic way does not lose value over time, but rather its aesthetic value is increased, leading to its renewal. The object that does not stagnate but keeps evolving, "assuming its character " over time and through use, is highly aesthetically sustainable. Creating an object with the vision that it should be able to evolve its expression and age in an aesthetic way perhaps involves fixating or "freezing" its expression at a point when it is not entirely finished.

Additionally, the time of becoming should be apparent, so that the recipient will be able to understand the past of the object or to familiarize herself with the process that precedes its current form and consequently internalize it—similar to how one's affection for another person tends to grow as one gains insight into that person's past.

Having the time of becoming manifest as visible or tactile traces in the object can also be a way to avoid the object seeming too "polished,"[3] disinviting use. In the case of polished objects, normal use tends to leave obvious signs of wear and tear very quickly, which is not flattering. But the wear and tear of object surfaces that have been designed to be touched and used can easily make the object appear more beautiful, more interesting, or more attractive. This form of wear—which merely (constantly) shifts the moment of fixation, and which thus supplements the story about the object—adds a permanent sense of renewal to the object that can help to maintain a strong bond between the thing and its owner. Another point to raise in this regard is that if an object is a frozen or fixated event, one could as well have fixed the event at some earlier or later stage, giving the object an entirely different expression.

For Ørskov, the artist (or the designer) should be led by the material in order to avoid beginning the design process with a set of restrictive conceptual thoughts about the design. By already having an idea of the finished design at the outset, one blocks off different potentially creative *flow* energies. The notion that the material should lead the process contains an element of aesthetic chance. Leaving the design to chance, to some extent, can be a way of surrendering to the process and the time of becoming, thereby making room for the creation of an object or product infused with the story of its becoming, instead of merely imitating the traces of the design process, conceptually.

Aside from appearing in the object's form, surface, or structure—in the guise of, for instance, obvious stitching, hand-painted elements, appliqués, or similar "fingerprints"—the time of becoming can comprise the story of the object or product. For example, a linguistic message can be attached to the product as a story about collaborating with another designer or an artist, or about how special, maybe local, craft techniques have been used to create it. It could also be done by giving the product a title that perhaps indicates how the material was shaped to fit the form. This way of charging a product with the time of becoming appeals in a lesser manner to a purely corporeal, sensuous detection of the process traces; nevertheless, such an approach can help create a strong bond between subject and object, as it invites users/recipients "inside," introducing them to so-called *insider* knowledge about what goes on "behind the scenes."

Making one's recipient party to part of the process that has preceded the moment when the object is ready to be "sent" out into the world—and thus involving her in the time of becoming—is a way of establishing a durable bond between subject and object. As such, it is a way that designers can endeavor to create an aesthetically sustainable expression. The traces of the time of becoming can add multiple layers of complexity to the object, helping to ensure that the recipient will not "finish" with it after a short period of time, but will rather become fascinated by it and want to return to it over and over. Consequently, a bond is formed between subject and object as a result of how long an object has existed for.

This leads me to Ørskov's second category of time: the time of existence.

## The time of existence

> In fact, she could never resign herself to buying anything from which one could not derive an intellectual profit, and especially that which beautiful things afford us by teaching us to seek our pleasure elsewhere than in the satis-factions of material comfort and vanity. Even when she had to make someone a present of the kind called "useful," when she had to give an armchair, silver-ware, a walking stick, she looked for "old" ones, as though, now that long desuetude had effaced their character of usefulness, they would appear more disposed to tell us about the life of people of other times than to serve the needs of our own life.
>
> (Proust, 2004: 90)

In this passage from Marcel Proust's *In Search of Lost Time*, the narrator describes how beautiful things can teach us to experience joy from sources other than "the satisfactions of material comfort and vanity."

Beautiful or aesthetically nourishing things can subsume and "fill" us in such a way as to seem spiritual, or at least to seem neither strictly physical nor reflexive. Aesthetically nourishing things satisfy in a very different way than material comfort or products that support our identity and status. Rather, such nourishing things fill us mentally while also *sustaining* us sensuously.

In Chapter 6, "The value of aesthetic sustainability," I will return to the concept of aesthetic nourishment.

Apart from verbalizing the notion of aesthetic nourishment, the Proust passage contains a description of the joy and love of things that are "disposed" to tell us about something, or that carry stories (about the past, for instance), and this joy is highly relevant when describing defining *the time of existence*. A thing can func-tion in ways other than "simply" to satisfy its immediate purpose—as, for instance, a sitting-machine; a keeping-the-cold-away-from-the-body-machine; or a pour-ing-food-into-a-bowl-machine, which would be in keeping with Bauhaus architect Breuer's description of a chair as a sitting-machine. In lieu of being strictly practi-cal, the function of a thing can be to give the individual an aesthetic experience, as mentioned in the section on "The beautiful" in Chapter 1. This point, exactly, is important to emphasize in connection with the time of existence.

Ørskov defines the time of existence in the following way: "The time of existence regards the duration that the object has undergone since its creation—its age, which is perceived by the 'marks of time' impressed upon it; it ages, fades, crumbles, or is consumed" (Ørskov, 1999: 85, transl.).

The time of existence has to do, then, with the signs of wear that characterize things which have existed for some time—maybe for a decade—and which have been used frequently. Things that have been used diligently are marked by this use, and they are often things that are or have been loved. In this way, signs of wear and disintegration testify to the loving bond that can exist between thing and owner; this bond is indicative of a special and lasting beauty.

Charging a design object with the time of existence in order to add aesthetic value to it, or to increase the possibility of creating a bond between the thing and the user, can be done in one or more of the following ways:

1   Striving to create objects that age gracefully or decay aesthetically by, for instance, ensuring that the material used can "contain" signs of wear without becoming uninteresting and unsightly; the fact that wear and tear leaves traces might even increase the aesthetic value of the object. This is a way of "setting the object free" or opening it up, letting it live its own life, and allowing this life to become a part of its expression. An option could also be that the object's form changes over time, that it expands or diminishes in some way, perhaps, to adapt to different needs. The possibility of changing the form could be another way of working with signs of use and change in the object, thereby integrating the time of existence into it—that is, if the signs appear clearly for the user to apprehend them.

2   Aiming to add "artificial" wear to an object or processing the materials in such a way as to render them "worn," or at least not entirely new—thus imitating or *mimicking* the wear that use and weathering normally would add—can be another way of charging an object with the time of existence. By creating an illusion of decay, the foundation is laid for the appearance of a bond between subject and object based on (the illusion of) history and use.

3   Integrating parts of recycled objects or materials and thus working with a mix of new and old elements can create an interesting expression that, as Ørskov might put it, consists of many different time sequences, among them the time of existence. For instance, one could integrate recycled leather, used buttons, patches, or wood into a new design object in order, partly, to reuse the traces that time has already left on the discarded objects, and, partly, to set the stage for new stories to emerge.

4   Working with *upcycling* products or objects—which entails updating objects that are no longer commensurate with the times due to, for example, an outdated style, an outmoded pattern, or faded colors—to emphasize the beauty of the already existing object by giving it a new "frame" or "packaging." Rather than erasing the traces of the previous lives of objects, *upcycling* therefore seeks to rejuvenate a previous aesthetic.

5   Creating objects that, in keeping with the wabi-sabi aesthetics,[4] do not appear too "polished," but which can accommodate wear and tear, as they do not seem new or "shiny" to begin with, is yet another way to introduce the time of existence to an object, hence infusing it with stories and feelings from a bygone time (even if this is an illusion).

The pivotal point in working with the time of existence is, in this context, focusing on the beauty of the imperfect and unpolished—the beauty of changeability; the fact that nothing is static but constantly moving and evolving; the dynamic quality of the variable expression; and the notion that the ravages of time can add depth and character to an object, rather than depriving it of its original qualities and characteristics.

The time of existence, just as the time of becoming, can establish a lasting bond between subject and object by adding to the durability of the object.

I want to end this section with another telling quotation from Proust's *In Search of Lost Time*, a quotation that underscores the fact that function is more than "just" practical function; aesthetic nourishment and aesthetic decay make things worthy of preservation and thus durable and sustainable:

> We could no longer keep count, at home, when my great-aunt wanted to draw up an indictment against my grandmother, of the armchairs she had presented to young couples engaged to be married or old married couples which, at the first attempt to make use of them, had immediately collapsed under the weight of one of the recipients. But my grandmother would have believed it petty to be overly concerned about the solidity of a piece of wood in which one could still distinguish a small flower, a smile, sometimes a lovely invention from the past.
>
> (Proust, 2004: 92)

## The time of being

Ørskov's phenomenological approach to encountering the world means that he considers detecting, or *taking in*, surrounding objects as a process wherein the subject physically experiences and interacts with them—touching them to sense their volume and dynamics and beholding them to apprehend their being. He describes the process as follows: "To experience (detect) an object is to be parallel to its time axis" (Ørskov 1999: 85).

The time of being is in many ways the most interesting—but also the most abstract—of Ørskov's concepts of time; it is the time it takes to *be-with* an object in order to detect it. The time of being can either be of long or short duration; conditioned by how complex the object, a kind of "inflection" takes place. But insofar as the time of being concerns the subject's *being-with* the object, this category is not as dependent on the thing-in-itself as is the case with the time of becoming and the time of existence. To a greater extent, the time of being is associated with the interplay between a subject and an object that the process of detection demands; the focus is the aesthetic experience itself.

As a designer, if one wants to bring into focus the aesthetic experience itself, one can strive to prolong the time of being with the aim of letting the recipient remain with the aesthetic experience for a sustained period of time, which Kant defines as the "chaotic stage."[5] Being kept in a state of chaos before realizing exactly what is happening corresponds to a particular kind of pleasure that is like the Pleasure of the Unfamiliar. Alternatively, one could choose to trigger an aesthetic experience in one's recipient that is characterized by a very short duration of the time of being—meaning that one's recipient is more or less able to detect the object (or concept) confronting her. Such an aesthetic experience will lead to the Pleasure of the Familiar.

A prolonged time of being can create a particularly durable bond between subject and object, as the experience of being in a state of chaos, momentarily, as well as the experience of overcoming that state—thus reaching an apprehension, a *taking in*, of the object—can be radically transformative.

The time of being can be prolonged by working with, for example:

- Complex compositions, asymmetrical formations, etc.
- A break from aesthetic universal principles for color harmonies by, for instance, challenging Itten's contrast by extension or one of his other color harmonies (Itten 1997).[6]
- Unconventional combinations or *sampling* of materials.
- Multifunctionality and flexibility.
- Charging objects with the time of becoming, or showing the process and referring directly (or perhaps indirectly) to the "hands" behind the objects (cf. the previous section on the time of becoming).
- Charging objects with the time of existence, or "making room" for signs of wear and decay to appear in the object—or imitating wear (cf. the previous section on the time of existence).

Thus, by working with complexity, unconventional combinations, *time charging*, or one of the other items above, it becomes possible to "force" recipients of a design object to stop and think about what confronts them; in this way, the time it takes to detect the object—and, consequently, to reveal the time of being—is prolonged. Prolonging the time of being betrays a potential for establishing a durable or sustainable bond between subject and object, as this temporal extension, and the *break* it introduces, "forces" the subject to experience a sense of presence. This kind of intense experience, following the experience of a prolonged time of being, tends to linger.

As Ørskov is a phenomenologist, his concepts of time are related to the sensuous, corporeal interaction with the objects that surround us, rather than with perceived values and symbols. This means that prolonging the time of being deals with a complexity of expression—the material, the form, the colors (see the above items)—rather than complex values and ambiguous signs.

Nevertheless, without "twisting" the concepts significantly, it is possible, similarly, to attempt to prolong the time of being as part of developing immaterial concepts or products that do not invite a sensuous experience.

The aesthetic experience—which is characterized by a prolonged time of being, and which is thus somewhat complex and "forces" the recipient to take a break from her daily routine—implies a form of self-experience. In being confronted by an object that breaks with conventions, the way things *usually* are, one's expectations of the object, and more broadly of the world, are likely to appear crystal clear. Hence, the object functions as a mirror, a mirror that seems to tell the subject that,

"The reason that you are struggling to detect my composition, my material, or my composition of elements is that I am challenging your expectations about the world and about things, generally."

The time it takes to detect the object, the time of being, consequently consists of a gradual insight into one's own basic assumptions. At the same time, the prolonging of the time of being involves the individual continuing on into the world with a new experience that will influence her future detecting of objects.

French phenomenologist Maurice Merleau-Ponty (1908–1961) describes the time of being (without using this term, however), or the time it takes to detect an object, as a dialogue between an experiencing subject and an object. He believes that the moods the subject becomes embroiled in as part of detecting, or engaging in a *subject–object dialogue*, condition preferences and taste:

> The things of the world are not simply neutral objects, which stand before us for our contemplation. Each one of them symbolises or recalls a particular way of behaving, provoking in us reactions, which are either favourable or unfavourable. This is why people's tastes, character, and the attitude they adopt to the world and to particular things can be deciphered from the objects with which they choose to surround themselves, their preferences for certain colours or the places where they like to go for walks.
>
> (Merleau-Ponty 2002: 48)

As part of the time of being, meaning the time we spend *with* the objects surrounding us, we "take in" the world. And herein lie the aesthetic experiences that condition our preferences, or that "teach" us what we appreciate most and consider the most "us."

I appreciate the idea that, as evinced by the quotation, the things we choose to surround ourselves with, the colors we prefer, and the places we like to walk are determined by what happens during the time of being—which is to say that in the *dialogue* between object and subject a kind of attachment takes hold, and that this lays the foundation for preferences and taste. Such an attachment and preferences can create a lasting or sustainable bond, based on aesthetics, between an experiencing individual and a thing or, perhaps, a place. The prolonging of the time of being—or what is implied by seeking intentionally to *expand* the time the recipient must spend in detecting and understanding a product—can thus be considered as a kind of appeal to letting oneself become preoccupied with apprehending the surrounding world and its objects. For, in doing so, and by virtue of the time invested, one is better positioned to engage in and attach to the things one considers worthwhile and beautiful.

In being alive to the world—sensing and interacting with the things around us, becoming absorbed in them, taking them in, spending time on and with them, enjoying them—we *recharge* our bodies, so to speak, with phenomenological wisdom. The more sensuously present we are, the more sensuously and aesthetically intelligent we become. With this as a point of departure, it becomes possible to develop one's sensuous, aesthetic intelligence by, for instance, "exposing" oneself to a number of tactile experiences that might make one sensuously wise to how surfaces feel. Or one could aim to stimulate oneself sensuously using color combinations that are culturally unfamiliar. As a designer, one can aim to stimulate one's recipient in a sensuous way, for instance, by employing diverse textures— or through "visual tactility" or the illusion of materiality in a two-dimensional medium—to create the foundation, accordingly, for a bond between the object and the recipient to form. Such a bond is based on the aesthetic experience of prolonging the time of being or on an expanded subject–object dialogue.

## Notes

1 See the sections on "Aesthetic decay, slow aesthetics" and "Sustainable thingness" in Chapter 3 for a further discussion about the value that decay and wear can add to an object or product.
2 See "Aesthetic decay, slow aesthetics" in Chapter 3.
3 See "Fleeting beauty" in Chapter 3
4 See "Wabi-sabi aesthetics" in Chapter 3.
5 See my description of Kant's tripartite ordering of the sublime aesthetic experience in the section on "The stages of the sublime" in Chapter 2.
6 See also the section on "Adhering to universal aesthetic principles" in Chapter 1.

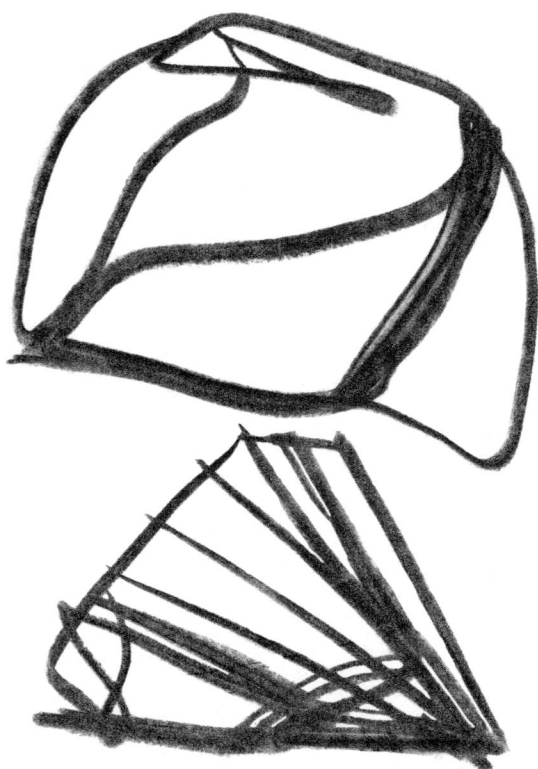

# 5 The magical thing

Some things are surrounded by an aura; they attract attention, compel interest, and establish an immediate and intuitive connection to the viewer—perhaps because they remind us of something or someone, or perhaps because they make us think about a particular place. Perhaps they remind us about who we used to be, who we want to be, or the part of our identity we value the most.

Such things are magical. They are not necessarily beautiful to look at, but they can ignite feelings and moods resulting in stimulating sense impressions or nourishing aesthetic experiences. Furthermore, they have the quality of making people want to spend time with them, wear them, or share the same space with them.

The notion of magical things is fascinating, the idea that things can contain some form of auratic power of attraction and spirituality, and that things can be charged with identity building or supporting elements that confirm who we are, or that remind us of who we want to be. This notion is associated with a certain kind of *animism*: "Animism is found, for example, in the idea that a person's spirit can be embedded in objects, or that objects in themselves can be equipped with a soul or an element of humanity" (Bille and Flohr Sørensen 2012: 23; transl.). The assumption is that there is no real separation between the spiritual and the physical world, that the thing-in-itself can be something.

The attention to magical things shifts the focus of the aesthetic experience from the experiencing individual to the experienced object, meaning it is the object that causes the aesthetic experience. This involves the thing-in-itself possessing aesthetic value, and that the aesthetic value is not solely determined, ontologically, by the interaction between subject and object, nor is it localizable to the subject's consciousness. At the same time, the focus on magical things requires that the designer actually has the ability to *fill* an object with the potential of giving the recipient an aesthetic, auratic experience.

It is not possible to completely disregard the importance of the recipient's (sensuous) interaction, cultural and social baggage, or individual realm of experience, as there can of course be no aesthetic *experience* without the presence of an experiencing subject! Nevertheless, I find that the notion of thing-magic or magical things can contribute a number of interesting perspectives on the concept of aesthetic sustainability.

My eight-year-old son owns many magical things, treasures that he keeps in small and large embellished boxes and chests. He takes good care of them, and brings them out often to look at and touch. This might lead one to think that these treasures consist of shiny, colorful thingamajigs or his latest toys—which is partly true. However, his most cherished belongings include odd things like a skeleton Lego figure that is missing its legs; a bashful pebble; several Lego blocks that appear rather "normal" and nondescript; a bedraggled feather of "bristly" barbs; a hair clip affixed with a peeling flower, which I think he found in the street; coral waste from a beach in Bali; and a plastic watch I bought for him from a street vendor in Marrakech. This collection of rather strange things is interesting, for why does he consider them to be treasures? Every time I ask him, he struggles to verbalize their importance. But what he keeps telling me—about the value of his skeleton Lego figure, his feather, his pebble, and his watch—is that "they are rare." Precisely the fact that they are "rare" is valuable to him. They are valuable in part for the simple fact that he only possesses one of each, and that he doesn't know any other children who have anything like them (for instance, in the case of the legless Lego figure), but they are also valuable in part because they contain emotional and aesthetic elements that combine to tell the story about how he acquired them—where he was when he found them, how they came into his possession, or who gave them to him.

The aesthetic, sensuous value of the objects, for my son, concerns the tactile pleasure of touching them and the fact that they contain a particular, *rare* beauty. He is fond of letting his fingers lightly brush across the barbs of the feather, even though, with each stroke, they come apart slightly more; he never tires of holding the pebble or touching the colorful Lego diamonds that form an essential part of the treasure. The sensuousness that amounts to the value of the treasure contains a high degree of tactility, and this tactility creates an intimate bond between my son and the things. At frequent intervals, he brings them to his lips to better sense their surface. Perhaps this is because lips consist of many nerve endings, and that by holding the things to his lips he is better able to experience their contours. But there is also a degree of infantility related to investigating things with one's mouth; my youngest son, who is one year old, puts everything into his mouth, a common practice, of course, for children of that age. I have a feeling that investigating the contours and surfaces of his favorite things reminds my eight-year-old of the safety and sensuousness (and adventurous spirit!) of being a toddler.

The beauty that characterizes my son's beloved things is crooked, rustic, and irregular—determined by traces of use and imperfections. As an example, one morning he found a pair of Lego legs that, strictly speaking, might very well fit the aforementioned skeleton figure, but after arguing with me, thrilled that the poor skeleton would no longer have to go legless, he reached the conclusion that the beloved figure simply did not seem *right* affixed with the newfound legs. The legs made it wrong, no longer *rare*, but rather too normal and all too similar to the Lego figures he owns.

Of course, there are things in his treasure troves that are fairly ordinary—for instance, a number of Lego pieces that do not appear different from the mountains of blocks that fill the baskets in his room. However, there is something about them, apparently, that makes them different, and, upon closer inspection, I think I know what it is: the Lego pieces selected from the baskets of "ordinary" blocks and elevated to the level of treasure each have a difference in nuance compared to the rest (this difference might be due to a factory error, in fact): some are transparent (not a very common feature, seemingly, of Lego pieces); others have been chosen based on a special function they served in regard to a vehicle or house he once built.

The emotional bond between my son and his treasures—his magical things—is strong. Each object represents something of great significance, and he uses them to make sense of the eight years he has been alive. Each might represent people he loves or places he has visited, but mostly they represent himself: his inner fantasy world, his way of thinking, and the games he plays. In other words, the objects are the essence of those moments of play when he reaches a *flow* state, forgetting about himself and becoming one with the act of playing, simply existing in the here and now, at peace with his surroundings.

Common to all the things—even the shiny, colorful, newly acquired—stored in my son's boxes and chests is the fact that they have no economic value: either they did not cost anything, or they were very cheap, and further, they have not necessarily been made from sustainable materials. Some of them, like the feather, are even coming apart. Their value, then, cannot be reduced to dollars and cents or to their material durability. Their value is rather of an emotional and aesthetic ilk.

The feeling is similar to when I'm at a flea market filled with treasure just waiting to be discovered, or when I'm in a vintage store swelling with wonderful garments and accessories. The rarest or most unique thing feels like the best find!

The above anecdote about my son's treasures might seem like a private and fairly subjective approach to understanding the concept of magical things. Nonetheless, despite the private tenor of the anecdote, analyzing it leads to a number of universal and generally applicable conclusions concerning what distinguishes magical things from ordinary or insignificant things. Regarding aesthetic value, we can thus determine that magical things are:

- "Rare," possessing a unique kind of beauty that is often rustic and irregular
- "Carriers" of stories and relations
- Nice to touch or sensuously stimulating (they make one want to revisit them over and over)
- Representative of the owner's (in the above example, my son's) identity and inner world or sphere of intimacy.

In my optic, it was especially interesting to discover that the things that seemed ordinary to me, but that my son cherished, represented *flow* moments to him—meaning moments of enhanced presence in regard to the activities he was involved in.

> Objects designed to appeal to these needs (i.e. functional, social/positional goods) are often rapidly outdated and unsustainable. Beyond these "middle-level" needs, however, there are the higher needs such as aesthetic and spiritual

needs. Products conceived to refer to these can appeal to our highest potential and in doing so, the very factors that spur unsustainable practices in objects are overcome. In the one example of prayer beads, at least, we have a product that is inherently sustainable, more than simply functional, and ubiquitous. This example demonstrates that this combination is at least possible to achieve. The challenge is to see if it is possible in more common, everyday products.

(Walker 2007: 50)

Magical things are associated with both spirituality and ritualistic behavior, such as my son's repeated touching of his treasures to his lips and stroking of his fingers across the ragged barbs of the feather. People tend to treat magical things in very particular ways, with a certain kind of awe or respect. Magical things can comfort and satisfy us aesthetically; they can also saturate our minds, almost spiritually, with *aesthetic nourishment*.[1] Nevertheless, as Walker points out in the quote above, it is challenging to sow the seeds of being able to "appeal to our highest potential" in everyday products.

Something being challenging does not mean, however, that it is impossible to achieve. In fact, charging an object (everyday or not) with aesthetic value—in the sense that it speaks to fundamental and aesthetic human needs, and that it possesses the potential to "appeal to our highest potential"—is an interesting way of seeking to create aesthetically sustainable objects.

The following sections are based on the bullet points regarding magical things and their characteristics, which are derived from the anecdote about my son's treasures. I will touch upon the characteristics of magical things—the emotional bond between object and subject; memory and sentimental value; tactile stimulation; and the aesthetic value of receiving or inheriting something from a beloved person—as well as the affectionate value one attaches to things in one's possession.

Additionally, I will briefly introduce the concept of *aura*, as a starting point for the rest of the book, and return to a foundational question of this book: what would it take to be seduced by—fall back in love, become fascinated all over again, with—a familiar object over a period of many years? Or, how can a familiar object become re-enchanted?

## The concept of aura

In his seminal text from 1936, "The Work of Art in the Age of Mechanical Reproduction," German philosopher and poet Walter Benjamin (1892–1940) describes the decay of the aura in the arts, and he associates this with how the means of reproduction seek to proliferate works of art (Benjamin 2007: 223–24). The decay of the aura is a decline of so-called "cult value" (ibid. 224), which typifies the classical work of art and which depends on its unique and non-reproducible character. However, the decay of the aura *only* pertains to cult value. Moreover, Benjamin travels in another type of aura, termed "profane," which is associated with the "secular cult of beauty" (ibid. 224). The experience of profane aura can be described as a *greater experience* of the physical world, or as a kind of divine experience, which cannot be equated with religious experience, nevertheless, as it does not imply "seeing" God. This divine, yet secular, experience involves a thingly dimension, and is thus connected to the realm of the everyday (Jørgensen 1990: 41). This relates precisely to the description and analysis of magical things in the previous chapter, as well as to Walker's theory about the "higher needs such as aesthetic and spiritual needs" of human beings As a designer, striving to accommodate these aesthetic and spiritual needs can lead to aesthetically sustainable products.

The experience of the profane aura constitutes the experience of transcendence or spirituality in the immediate world of the everyday that surrounds modern human beings. It is an experience of something distant or "divine" about the immediate world of experience. The profane aura is the appearing of the transcendental in an immediate, fixed *here and now* (Jørgensen 1990: 41).

Benjamin celebrates the profane aura and does not consider the loss of cult value as a negative development. The experience of the profane aura, Benjamin says, can be found in Baudelaire's poetry, for instance, which deals allegorically with the experience of the modern metropolis (Baudelaire 1998: 167–211). According to Baudelaire, modern allegorical poetry is precisely characterized by the concurrence or convergence of eternal and relative elements, which likewise describes the Benjaminian aura.

Moreover, the experience of profane aura is associated with the sublime experience, which in a similar way is marked by a sudden insight into something immeasurably great or distant—an insight which can be defined as the culmination of the process, or three stages, that Kant uses to classify the sublime: 1) the encounter with the triggering phenomenon; 2) the paralyzing interval; and 3) the culmination of the experience, which involves an elevated state of calm.[2] The sublime experience, in my optic, can very well be triggered by a sensuous phenomenon or by a (magical) thing. Both the sublime experience and the experience of the profane aura are characterized by a dialectical movement, or a form of dialogue, between the finite and the infinite; as part of the experience, one is both extremely present and *beyond* time and space. In that sense, the experience of the profane aura can seem similar to the particular kind of presence associated with being in a state of *flow*, which I described in relation to the anecdote about my son's treasures as a crucial factor of value creation.

Experiencing the aura of the magical thing can send the individual into "time travel" that involves a momentary fusing of past and present, near and distant. For an instant, "has been" and "here and now" melt together, and a passage between past and present is carved out. This kind of experience can lead to insights concerning past experiences and feelings, and thus can put the individual's present into perspective. Profane aura experiences are determinate and insightful, and may contribute to creating a sense of coherence in life.

## The re-enchantment of familiar objects

The last time I experienced this was on the commuter train between Stockholm and Gnesta a few months earlier. The scene outside the window was a sea of white, the sky was gray and damp, we were going through an industrial area, empty railway cars, gas tanks, factories, everything was white and gray, and the sun was setting in the west, the red rays fading into the mist, and the train in which I was traveling was not one of the rickety, old, run-down units that usually serviced this route, but brand-new, polished and shiny, the seat was new, it smelled new, the doors in front of me opened and closed without friction, and I wasn't thinking about anything in particular, just staring at the burning red ball in the sky and the pleasure that suffused me was so sharp and came with such intensity that it was indistinguishable from pain. What I experienced seemed to me to be of enormous significance. Enormous significance. When the moment had passed the feeling of significance did not diminish, but all of a sudden it became hard to place: exactly what was significant? And why? A train, an industrial area, sun, mist?

(Knausgaard 2012: 306–07)

This passage from volume one of Knausgaard's novel *My Struggle* (a work that, like the remaining five volumes, particularly volume six, contains a plethora of well-described accounts of aesthetic experiences) can be said to verbalize an experience of the profane aura. Most will probably recognize the feeling of suddenly seeing familiar objects or surroundings in an altogether new way, of suddenly feeling overwhelmed by the beauty of everyday surroundings or objects, which are not otherwise special or spectacular in any way. When suddenly overwhelmed by beauty in this manner, it can feel so wonderful as to almost be painful; at the same time, it can seem highly significant despite the fact that the reason of the significance can be difficult to explain. The feeling of having experienced or witnessed something special will often stay with the individual for a long time following the event. The event is far-reaching, so to speak.

Having an eye for beauty and feeling aesthetically *nourished* by one's surroundings and the objects found here, no matter how ostensibly insignificant they might appear, in many ways corresponds to what Jonathan Chapman defines as "normalcy" (Chapman 2011: 63).[3] This concept reminds me of my son's treasured thingamajigs, which, on the whole, I find insignificant and quite commonplace, as they seem similar to a host of other bits and pieces in his possession. As mentioned, my son attaches a magical quality to these otherwise common, everyday objects: they all remind him, sensually and emotionally, of moments when he would have been entirely "lost" in playing, thereby experiencing a sense of enchantment by being at once highly present and outside of time and place. Letting his fingers search the surface of ostensibly common or insignificant things, his sense of enchantment is reactivated for a while. The otherwise nondescript doodads compose, in some way, a portal to a satisfying, sensuous state, wherein the surroundings and the subject (in this case, my son) merge to become one. Normalcy is a positive concept, as, precisely, in the familiar, in the repetition, and in the normal—or in the *re*-fascination and *re*-enchantment, the recurring point of seduction—there is a very particular kind of extended (or *slow*) aesthetic pleasure—a durable, and thus sustainable, pleasure.

However, it takes a particular frame of mind for the recurring seduction, the re-enchantment, or the experience of beauty in one's usual surroundings to come about. Only through having an open mind—through being a particularly sensuously present person—is it possible to experience, and to derive nourishment from, this wonderful, almost painful, kind of beauty. Perhaps this is what Friedrich Schiller had in mind when he wrote that being receptive to beautiful and sublime experiences is innate to every human being, but realized that this potential can only happen through art or through aesthetic education, which is liberating.[4] The possibility of being open and receptive to affective aesthetic experiences is innate to all, but not everybody will manage to liberate and thus realize this potential. For Schiller, this potential equates to an experience of the sublime: the soul-shocking aesthetic experience that rouses human beings from the slumber which had been preventing the aesthetic, the liberating, and the insightful from blooming. Why? Because the phenomenon or the object, with the potential to trigger a sublime experience, aggressively intrudes upon the subject's attention to such an extent that it cannot be ignored; in a manner of speaking, it "twists" the arm of the subject.

In particular, it is in breaking with the trivial that it becomes possible to have an aesthetic, or painfully touching and beautiful, experience. The trivial, or repetition without *nerve* and soul, is soporific and is unconnected with the positive concept of normalcy described above. The trivial is unrelated to the everyday magic that Knausgaard expresses in *My Struggle*, a novel which contains a number of passages dealing with draining triviality. The *re*-enchantment and *re*-fascination that the *rewarding* extension of time and the power of repetition (normalcy) involve presuppose a degree of presence and nearness. In particular, repetition can be highly pleasurable, which is the case in the Pleasure of the Familiar and the Pleasure of the Unfamiliar, equally. Repetition can function as the seat of durability. The trivial and the soporifically mundane, on the contrary, are the enemy of any form of beautiful experience. Experiencing beauty in mundane objects and activities requires an openness, a presence, and an unfiltered or pre-rational approach that goes beyond the merely trivial.

In *Detecting Objects*, Ørskov writes that we have both a utilitarian and a conceptual relationship to most everyday objects. However, considering the everyday objects of our surroundings on their terms, existing alongside us in physical space, and hence not just as objects of use, the objects will reveal an "independent existence" (Ørskov 1999: 76). Ørskov calls this approach a sensuous and emotional detection of objects. Such a form of detection, or apprehension, can lead to an insight or an understanding about our surroundings and the objects we may encounter that goes beyond or, at the very least, represents an alternative way of looking at the world not afforded by rational or thought-based approaches.

Thus, this alternative (sensuous, emotional) way of detecting the world can reveal something other than cognitive decoding; as such, it is similar to what I have termed *pre-rational* experience above. It is not just anti-rational or irrational; in many ways, it can be described as what *precedes* the rational, the most fundamental and human approach to understanding and acting in the world. For this reason, namely, wishing to understand Ørskov's approach to the world, it can be fruitful to observe children and their way of relating to objects and spaces (apropos of the anecdote about my son's treasures or magical things). The naive or immediate approach to the world, which defines how children experience it, characterizes the sensuous and emotional detection of the world.

The sensuous, emotional detecting of the world, however, can be difficult to hold onto or submit to fully after having been used to encountering the world primarily through one's rational, analytical (critical) sense. In a sense, the self has to shut down its desire to systematize and conceptualize its surroundings in order for the frame of mind that will allow the sensuous, emotional, and pre-rational detecting of the world to appear. Nevertheless, it is precisely this alternative approach to looking at the world and objects that forms the basis for a powerful experience of beauty and for a re-enchantment of the familiar. Moreover, this approach contains the potential for removing that which is trivial—if not forever, then at least for a (painfully beautiful) moment.

What is particularly interesting in this regard is whether or not the designer can contribute to "waking" the recipient from the soporific state of the trivial by infusing things or products with aesthetic value, or if a design object can place the recipient in that state or frame of mind which is a condition for the pre-rational, emotional, and sensuous detection of the world.

As mentioned, the sublime aesthetic experience, according to Schiller, has the ability to "wake" the subject and thereby, momentarily, make her receptive to a powerful experience of beauty. Likewise, in his aesthetico-philosophical text "The Sublime and the Avant-Garde," Lyotard describes the sublime aesthetic experience as that "space" where presence is carried through. Furthermore, he directs artists to strive to provide recipients with an experience of sublimity, which amounts to a form of strategic planning as regards the aesthetic experience. In relation to clarifying the content of aesthetic sustainability, his reflections on the subject are therefore highly relevant to consider (Lyotard 1991: 89–107, 135–43). Surprising, shocking combinations (regarding forms, materials, and colors—or concerning signs or symbols), for Lyotard, can be an effective way of making the recipient "sit up" or breaking someone's "hypnotic" everyday dealings.

I will return to the planning of the aesthetic experience in Chapter 7, "Aesthetic strategy," where I will also address Lyotard's text in more detail. At the present stage, however, it is interesting to note how Schiller as well as Ørskov and Lyotard equally emphasize the importance of cultivating a sensuous, emotional frame of experience, compelling the subject to become present to the experience of the beauty of the proximate, the familiar, or the usual. Additionally, according to Lyotard, it is crucial for artists to consider it their main task to experiment and combine elements in new and shocking ways, since works appearing as the result of such a process are particularly apt to rouse recipients from the slumber of trivial existence. The time of becoming can thus have an effect on whether or not a work or an object is able to "touch" recipients, momentarily removing all filters and making them open to the beauty of the world.

### Turning things upside down

By literally turning things upside down and ripping them from their context, Marcel Duchamp (1887–1968), artist of the ready-made, sought at the beginning of the 20th century to rouse modern human beings from the grip of their trivial lives. In the creation of ready-mades, making viewers consider familiar objects in new combinations, Duchamp removes the most trivial and unsightly objects (e.g., a urinal or a bicycle wheel) from their everyday context, refashioning them as art pieces; in this way, removed from their usual context, they suddenly appear different and altogether anti-functional. No longer is the focus on use and function. The objects are now merely forms and materials in a space, and the experiencing subject must consider the apparently familiar things anew, with what might be called a purified frame of mind, or a *beginner's mind*. In this way, the time of existence—meaning, the time it takes for the viewer to detect and take in the object[5]—is extended. In addition to being torn from their context, Duchamp's objects are often literally turned on their head, and the viewer is thus forced to look at these articles for everyday use in a new light: their forms, colors, and material composition are emphasized, whereby they gain aesthetic value all of a sudden. Perhaps it is revealed how dynamic and malleable their forms are, or perhaps their material composition appears more interesting than first assumed. Displaying articles for everyday use *re-enchants* the everyday, so to speak, as well as what has already been de-enchanted (Jørgensen 2001: 343), thereby reflecting an experience of profane aura—an experience of something magical occurring within the confines of everyday life.

In a similar way, and very like Lyotard's guidelines for how artists are to "wake up" their recipients, the surrealist artist Meret Oppenheim (1913–1985) draws upon surprising, strange, shocking combinations in her *Object* from 1936. The *Object* consists of a cup, a saucer, and a spoon clad in fur. This work speaks to the tactile sense in that just by looking at it, it seems clear how it would feel to touch the fur-covered cup to one's lips: wrong and abhorrent, but also fascinating. The combination of cup and fur is unheard of; a cup is supposed to be clean, smooth, and cool (unless it contains a hot liquid), but the addition of fur makes it anything but. Nevertheless, by forcing the recipient to consider her own cultural framework—including expectations, connotations, and habits—she is momentarily torn from her habitual, trivial dealings with everyday objects, and in the break thus introduced, she becomes sensuously present and open to an experience of beauty.

However, are designers similarly able to help their recipients rediscover familiar objects or discover what is "rare," unique, or special about a familiar object or surroundings? As a design by definition should have some kind of function, designers cannot incorporate the same kind of anti-functionality into their works as Duchamp and Oppenheim; yet, it might still be possible to "borrow" the intention of the ready-made artists and the surrealists to force recipients to see the familiar in a new light, and thus carry through an openness to see the beautiful in the things we take for granted and do not typically regard in any particular way.

*Tactile spirituality*

The writings and works of Russian artist and philosopher of art Wassily Kandinsky (1866–1944) contain a form of magical experience of the world. In his art and philosophy, Kandinsky insists that there is a trace of *spirituality* in the immediately sensuous world of phenomena and in our daily surroundings, and the aim of his art is to express the almost religious feeling that life and the world, and all its objects, fill him with (Long 1980: 47). As such, it is the experience of beauty and the spirituality of the familiar that visual art should express, according to Kandinsky. It is precisely the experience of everyday spirituality, which is also expressed in the quotation from Knausgaard above, that Kandinsky favors.

In his aesthetico-philosophical work *Concerning the Spiritual in Art*, Kandinsky introduces an interesting concept that he calls "the guiding principle of inner need" (Kandinsky 2008: 62). This principle is based on three elements: the personal element, or the artist's ability to express the particular characteristics of her personality; the temporal element, meaning the ability of the artwork to capture the essence of its period; and the spiritual element, which concerns the ability of the artist to express what is specific to art, that which is pure and everlasting, existing beyond the confines of time and space (Kandinsky 2008: 74–75). All three elements are essential, but, were the third element to carry greater weight than the other two, Kandinsky would consider this a sign of greatness in the artwork and the artist, equally: "this is the element of pure artistry, which is constant in all ages and among all nationalities. [ . . . ] Only the third element—that of pure artistry— will remain for ever" (Kandinsky 2008: 75).

Kandinsky's insistence that our familiar, everyday surroundings contain traces of spirituality is an interesting point to consider further. The notion of sensuous spirituality is similar to the idea of profane aura, which, again, is related to the question regarding when an object can be considered aesthetically sustainable.

But is it possible to translate Kandinsky's three principles concerning the inner need into a set of guidelines that could be used in connection with the creation of aesthetically sustainable design?

The first and the second elements—the personal element and the temporal element—can easily and suitably be "translated" into permanence and variation, respectively—or into repetition and renewal—which I described as being a characteristic of the durable, sustainable aesthetic expression in the section on "Aesthetic decay, slow aesthetics" in Chapter 3. By this, I mean that the durable object possesses a kind of permanence marked by recognizability—that the designer behind the object is recognizable (the designer's signature, so to speak), or that the object triggers the Pleasure of the Familiar, assisting the recipient in quickly detecting/ decoding and using the object. Furthermore, this means that the object experienced as both interesting and attractive at the same time exudes a certain degree of variation or a renewing, contemporary element. This should not be taken to mean that the object's expression adheres to common trends, but rather that its expression appeals to governing basic assumptions or myths (e.g. the tendency toward *slowness*).

However, the third element, which constitutes a vital part of the principle of inner need, is somewhat "unruly." The artist, writes Kandinsky, must express something that exists outside of space and time. But what does that mean? And can this kind of principle be in the creation of design objects?

Kandinsky goes on to write that the principle of inner need involves a revelation about "an ever-advancing expression of the eternal and objective in the terms of the periodic and subjective" (Kandinsky 2008: 77). This might sound fairly abstract, yet it is exactly similar to what Knausgaard experiences travelling by train through an altogether ordinary industrial area on an altogether ordinary day: a feeling of overwhelming beauty, which transports him outside of the periodic and the spatial, for a brief moment. Additionally, Kandinsky's description of aesthetic revelation shares many similarities with the enchantment of normal, unassuming, familiar objects that my son picks up on when giving himself over completely to the forget-ful magic of play—the near and the distant become as one. In other words, my son experiences the "expression of the eternal and objective" during moments of play and in being (repeatedly) fascinated with his treasures—treasures in the shape of modest little things, even if many of them are *rare* or unique or possess a certain "wryness", which makes them different, or which have somehow managed to elicit my son's interest in them (maintaining his interest for an extended period of time).

Precisely at this point—in understanding the part elements that make apparently ordinary items or surroundings seem special somehow or that make the owner of a thing form an emotional attachment to it—is the key to being able to plan, strate-gically, the integration of similar elements into the creation of new products and objects. This is also the key to creating durable, lasting bonds between things and people, and to creating aesthetically sustainable objects.

## The lover's thing

> Objects capable of sustaining long-lasting relationships with consumers are rare. Most emotional attachments are withdrawn once the honeymoon period draws to a close.
>
> (Chapman 2011: 66)

I have previously mentioned and made use of the *honeymoon* metaphor that Jonathan Chapman employs in his book *Emotionally Durable Design* to describe the immediate fascination with newer (shinier, attractive) things, which in time will lose their allure, becoming obsolete. In this section, I will discuss whether or not the honeymoon period can develop into a sustainable, loving relationship and what this would then require. Moreover, I will consider the aura or radiance that can encircle those things that, as human beings, we own, love, and cherish.

Several authors have written about and celebrate the magical things and clothing pieces that belong to, or have been worn in the presence of, the beloved. In J.W. von Goethe's work about the great (unhappy) love from 1774, *The Sorrows of Young Werther*, the main character creates an almost auratic radiance around the blue coat and the yellow vest he wore the first time he danced with the unattainable object of his love, Charlotte. Every time he wore the pieces after that, he would recreate the feeling of the magical first dance, thereby creating a brief passage to the past. The blue coat and the yellow vest blend the distant (and romanticized) "then" with the throbbing (painful, lonely) "now." It is thus not a surprise that he should be wearing these when he kills himself.

Roland Barthes' philosophico-poetic work from 1977, *A Lover's Discourse*, displays the torrent of speech of a man or woman in love; as part of the speech, the beloved's body and clothes are objectified and fetishized. As a result, the beloved's pieces of clothing assume a level of mystical, or almost religious or spiritual, provenance. In a sense, they carry something of the beloved's identity; they are a part of him or her and therefore magical. They are more than just (meaningless) things.

With *A Lover's Discourse*, Barthes does not want to explain love, tell *about* love, or establish a psychological portrait of the person caught in love. Rather, the fragments of the text create immediate images of love—and the attenuating feelings and moods—that any subject-in-love would be able to recognize: "At every moment of the encounter, I discover in the other another myself: *You like this? So do I! You don't like that? Neither do I!*" (Barthes 2001: 199).

The things that are associated with or bound to the person—or the people—one loves have a very particular meaning. But not all of their belongings, of course! Some of their things in particular will appear magical or auratic—those things can be considered the essence of one's object of affection; in a way, they *are* that person. A high degree of tactility is linked to such things. For instance, it could be a sweater that seems to always carry the scent, ever so slightly, of the beloved, and the surface texture of the sweater might give the feeling of being close to the beloved. It could also be a notebook containing the beloved's handwriting, which, because of the uneven texture created by pen against paper, can function as a sensuous portal into the beloved's mind. Things possessing such qualities are central to the concept of aesthetic sustainability; they are carriers of stories and relations—and of magic and attraction; they have the power to erase the borders of distance and time.

With deep-felt love, which is supposed to replace the intoxicating infatuation that belongs to the honeymoon stage, follows a certain fascination with the beloved's common or unoriginal characteristics and thus an acceptance of the beloved's true "self." Based on this is an acceptance of the fact that the beloved's identity consists of something unique, of course, but also of something common and *human*. Such a point of fascination is in sharp contrast to the disappointment following the insight that the beloved is just one among many potential love objects; this fascination, thus, is the opposite of what could be called the *serial honeymooner's* craving the rush and newness of falling in love.

The young Werther is in many ways a serial honeymooner. To a great extent, he is in love with the image of Charlotte he has created and the fact of being in love, rather than Charlotte herself:

> Wounded by a remark he overhears, Werther suddenly sees Charlotte in the guise of a gossip, he includes her within the group of her companions with whom she is chattering (she is no longer the other, but one among others), and then says disdainfully: "my good little women" *(meine Weibchen).*

A *blasphemy* abruptly rises to the subject's lips and disrespectfully explodes the lover's benediction.

(Barthes 2001: 28)

In the novel *Identity* by Czech writer Milan Kundera (b. 1929), acceptance of *and* fascination with the beloved's ordinariness are expressed in a situation when the man of the relationship finds some letters hidden beneath his beloved's underwear:

> He leaned into the open wardrobe, staring at the brassieres, and suddenly, without knowing how it came about, he was moved. Moved in the face of this immemorial action of women hiding a letter among their undergarments, this action by which the unique and inimitable Chantal takes her place in the endless procession of her peers.

(Kundera 1998: 32)

The man in love reconciles himself to the beloved's true identity by accepting her humanity—that is to say, by accepting that she is simply a human being, who, in her humanity, is like everybody else. That which is human is ordinary and everyday-like, but this is everything but negatively understood. The everyday is auratic in the sense that it turns into something perpetually rewarding, fascinating, and durable.

Exactly the union and reconciliation of the beloved's humanity and normalcy remind me of a novel I read some years ago: *The Museum of Innocence* from 2009 by Turkish author Orhan Pamuk (b. 1952). In the novel, the main character, Kemal, spends most of his life collecting things that belong, or have belonged to, or that in some way remind him of, his beloved Füsun, whom he, for many different reasons, did not marry in time, thereby ruining his chances with her. The different things he collects—cigarette butts, an old ruler (which he sometimes cannot even stop himself from *tasting*), hairpins, several garments (which feel and smell like her), and small porcelain figurines from her parents' home—afford him auratic experience. When touching, smelling, or tasting these things, he creates a passage from the unbearable present to the wonderful, albeit lost, past. For a moment, the distance between "then" and "now" is obliterated: Kemal undertakes a brief travel through time. He recreates the happiest moments in his life through the things he has collected, and he maintains his fascination with Füsun on the strength of her ordinary mortality; the things, in fact, emphasize her vulnerability and humanity, which is specifically what he finds so moving.

In 2012, Pamuk opened a museum in Istanbul with the same name as his novel, *The Museum of Innocence*. The museum displays all the things that Kemal collects over the course of the many pages of the novel; the exhibition must be considered the ultimate collection of trivial items and common doodads—or, perhaps rather, the ultimate collection of extremely personal "treasures." In conjunction with the opening of the museum, Pamuk was interviewed by *The New York Times Style Magazine*:

> Let us say in the pocket of one of my old coats I find a movie ticket from many years ago. [. . .] Once I see the ticket, not only do I remember that I saw this

movie, but also scenes from this movie, which I think I have entirely forgotten, come back to me. Objects have this power, and I like it.

(Brubach 2012)

As Pamuk puts it, what is interesting about these specific things is the fact that they have the power to create a passageway between past and present, through which forgotten experience and moods can "travel." In other words, things can be carriers of events and emotions.

The love of the beloved's humanity and normalcy can represent a relationship to things which not only fascinate by virtue of their news value, but which maintain the recipient's or the user's interest for years. As the opening quotation of this chapter points out, "Most emotional attachments are withdrawn once the honeymoon period draws to a close" (Chapman 2011: 66). But some things stand out. At times, a strong bond can form between an object and a subject that is based on aesthetics or on emotions and, largely, on an appreciation of the thing's normalcy or ability to evoke pleasure time and again. There are certain things that one simply will never tire of, or at least not until after a long while, maybe not even for a number of years, nor will one ever tire of the pleasure they evoke based on continual and sensuous use.

### Designing magical things

By implementing into one's designs some of the qualities that characterize magical things, making them carriers to be filled with subjective emotions and stories, it becomes possible to create the foundation for a strong emotional, and sustainable, bond between object and subject—sustainable in the sense that the recipient would be loath to replace or get rid of an object that has the potential to open up passageways to earlier events or to a beloved person, or that is able to unite proximity and distance.

Is it possible to create objects or products that can affect a broad target group in such a way? The events that make up a life and the people one meets and falls in love with along the way are associated with the subjective, individual life, and so of course to one's unique journey through life. Therefore, on the face of it, it seems impossible to draw upon something so particular, singular, and personal. Yet, it is interesting to note that the objects that manage to erase time and distance— thereby creating a connection to earlier times and beloved people, if only for a short while—have not a few characteristics in common with meaningful events and people in a person's life.

In *The Museum of Innocence*, the everyday is re-enchanted, as the many trivial objects collected by the main character through the years are removed from their context and placed into a new one, the museum (apropos the previous section about ready-made art pieces and their ability to re-enchant the everyday). In this way, these objects are elevated to something more than trivial, idiosyncratic things. The collection celebrates the repetition and the normal; the main character's collection, among other things, consists of a plethora of cigarette butts and hairpins, which materialize the everyday routines of his beloved—routines that make her exactly like everybody else, humanize her. Human, but not trivial! It is namely in the human and in the everyday normalcy that we find the exotic and the fascinating. Or, it is through the repetitive, insistent, almost ritualistic safeguarding of the everyday and its routines, shaping a life out of different events, that the main character creates a sense of durability and value. Subjective temporality—in the form of plethoric, concrete, apparently insignificant items—enables a revelation of the eternal and objective,[6] which can eliminate time and space momentarily, making the ostensibly extremely personal and private collection of things and cigarette butts interesting and meaningful for many others than the love-struck main character in the novel. More than a collection of things, it comes to represent or celebrate persistence, repetition, and humanity. As described in the section on "The re-enchantment of familiar objects," *re*-fascination and *re*-enchantment compel a recurring seduction of the subject that leads to a highly particular durability and a sustained (or *slow*) aesthetic, sensuous pleasure.

> When chronological time comes together with the eternal now, a different understanding of time results, that of sacred time. Sacred time is the cycle of time. [. . .] In contrast to the linear progression of chronological, secular time, sacred time presents us with a circular view of time, with repetitions and recurrences; it can be regarded as an endless repetition of eternities.
>
> (Walker 2007: 143)

In appreciating and maintaining some of the "normal" things we surround ourselves with, a confluence occurs between what Walker here calls *chronological time* and *sacred time*, resulting in a temporal extension. Designers should see the potential of appealing to consumers to consider repetitions and recurrences as something valuable, rather than as something trivial and uninteresting.

Based on a number of different individuals' statements about the ability of magical things to create passages through space and time or to "appeal to our highest potential" (Walker 2007: 50), I hypothesize that it is possible to draw up criteria for the characteristics of *lasting* things. Drawing on pattern recognition or general characteristics of things that function as "carriers" of relations and stories, or which have been charged with time, it is possible, in other words, to reach an understanding of what characterizes these things. Once such characteristics have been identified, designers can seek to establish durable bonds between subjects and objects when designing new objects.

In the vein of tragic poetry, designers can strive to provide recipients with a cathartic experience[7] by employing universally human thematic material to reach the broadest audience possible and to trigger a profanely auratic experience. The profane aura might take shape as:

1  The realm of childhood and the sensuousness belonging to it. This could be implemented by a variety of surfaces and by combining materials that can activate the tactile sense and thus put rationality into "stand-by" mode.

Or techniques can be used that connote childlike memories of "grandmother" in the form of lace, knitted work, and crocheted elements, all of which, furthermore, include the time of becoming as well as storytelling about the use and maintenance of old, local craft techniques.

2    The beloved's things and the celebration of normalcy or the magically human about the beloved, an element that is anything but trivial and could appear in the guise of everyday clothes or use items that "celebrate" the everyday, thereby exoticizing the normal.

3    The experience of *flow* or the experience of being one with or at home in the world, which find expression in objects imbued with repetitions and a regular and harmonious appearance symbolizing the feeling of "floating" along, unencumbered and highly present at the same time.

In Chapter 7, "Aesthetic strategy," I will explain further how designers can work strategically to achieve the intended recipient experience in order to establish a durable bond between subject and object.

## Notes

1 I will return to this concept in the section on "Aesthetic nourishment" in Chapter 6.
2 See my account of Kant's classification of the sublime aesthetic experience as consisting of three stages in Chapter 2.
3 See the section "Sustainable thingness" in Chapter 3, where I also use the concept of "normalcy."
4 For a mention of this, see Chapter 2.
5 See Chapter 4, "Designing the temporal object."
6 See Kandinsky and his guiding principle of inner need, which I mentioned and translated into the context of the design object and the design process in the section on "Tactile spirituality" earlier in this chapter.
7 Cf. Aristotle's view of catharsis, which is associated with the good, efficient tragedy that leaves the viewer "purified." See Chapter 2 for an in-depth discussion of catharsis in relation to the sublime.

# 6 The value of aesthetic sustainability

> Design can be understood in terms of the added value it brings to manufacturers and consumers and the role it plays in the market system of supply and demand. However, this economic evaluation provides only a limited grasp of the meaning of functional objects. It does not account for those desires, needs and consumer preferences that cannot be expressed in economic terms.
>
> (Walker 2007: 187–88)

As Walker suggests, value is more than an economic or functionalist term. When purchasing an object, "value for money" is less of a determining factor in assessing its overall value. Emotional needs, wants, and aesthetic preferences play a large part in the decision to buy a certain object; such needs, wants, and preferences are not rational parameters. Rather, they can be categorized as irrational or anti-rational—or perhaps even pre-rational, understood in the way that they don't cover intellectual calculations of "for or against" something.

Is it then possible, as a designer, to control consumers' irrational preferences or desire-based decisions? Further, might it be that irrational or pre-rational attraction to a given object is a universal factor, governing our relations to the object world?

The notion of value creation is traditionally primarily concerned with maximizing profit. Nevertheless, from a sustainability perspective, building value into a product has more to do with creating relations—to the local community, employees, and manufacturers, as well as to consumers. Moreover, an important aspect in the creation of sustainable value is longevity: manufacturing products that will last for years, both in terms of quality and functionality, in order to create long-lasting value for consumers. Aesthetically sustainable value is closely connected with sustainable value, but is, at the same time, permeated by the aesthetic experience. Aesthetically sustainable value is based on a product's ability to please the viewer/user with its expression for years to come.

According to the German philosopher Alexander Gottlieb Baumgarten (1714–1762)—who is considered to be the founder of aesthetics as a scholarly discipline—aesthetics can be defined as the process of attaining *insight*: aesthetics constitutes a sensing form of insight, *cognito sensitiva*, which is based on emotions and feelings. Aesthetic insight is something very different from logical or rational forms of knowing (Jørgensen 2001: 235), as it can be characterized as pre-reflective and pre-linguistic. Aesthetic insight has to do with the human subject's immediate, unfiltered encounter with the world. The potential of the aesthetic experience to generate insight into the workings of the world is thereby a trait common to all human beings. The aesthetic experience, experiencing beauty, has the ability to momentarily pull us out of our trivial and everyday lives. Everybody can have aesthetic experiences, which tend to be fairly similar regardless of the individual subject; or, rather, we cannot so much describe them as we can feel, sense, and perceive them in a similar manner, since it seems that when we are consumed by an aesthetic experience, words cease to dominate. The aesthetic experience is in a sense indescribable.

Whether or not the object of the experience is generalizable is open to debate, but it is precisely this question that is at the root of determining the nature of aesthetic sustainability. My hypothesis is that aesthetically sustainable value can be added to an object, and that this value can serve as the foundation for establishing a bond between subject and object, or between people and things. To a certain extent, aesthetic value comes from the object or the thing-in-itself[1] and partly from the interaction or dialogue between the object and the experiencing individual through the time of existence and the aesthetic experience. An object can be *charged* with aesthetically sustainable value, and the aesthetic experience can to some degree be planned and thereby *controlled* by the sender or the designer of an object or product (or concept).

As shown in Chapter 4, one way to add aesthetic value to an object is to charge it with the time of becoming or the time of existence—another way would be attempting to prolong the time of being.[2] Additionally, designers can seek to create an object that is neutral enough to enter into many different contexts, which would allow it to create pleasure for a great number of users. Multifunctional objects are yet another means of creating long-lasting aesthetic value; the ability of such objects to alter appearance and expression accommodates the changing needs of a person's different life phases.

Charging a product with aesthetic value is contingent on specific connotations. Individual designers can embed objects with different connotations, but it is important to realize that the aesthetic experience and the subsequent aesthetic and emotional attachment to an object are highly subjective—attachment to an object could be based on individual memories, for instance. It is therefore crucial to not "close off" or invest an object too heavily with connotative value, so as to render it resistant to personal meanings. Some connotations will naturally be embedded in the product during the design process, but it is important, concomitantly, to create a framework that will allow users to invest the object with their own personal histories and experiences. Designers may want to give users a certain aesthetic experience and to embed objects with specific kinds of aesthetic value, but it takes a high degree of audience awareness to be able to charge an object with the right elements that will allow it to be decoded or read in the intended way.

I will return to the strategic planning of the user experience in Chapter 7, "Aesthetic strategy," but for now I want to focus on the importance of communicating the inherent value of aesthetic sustainability. Unless consumers gain an understanding of the specific production and design processes, it can be difficult to accept that sustainable and durable products must cost more than their mass-produced counterparts, which will perish relatively quickly.

## Communicating sustainable aesthetic value

A cornerstone of working with sustainable design, and not least of all aesthetic sustainability, is convincing consumers of the importance of buying fewer but better, longer-lasting products—longer-lasting in terms of both quality and aesthetic expression. Communicating the value of spending money on well-made and well-designed products—which are also ethically produced, guaranteeing workers proper conditions—that will last for many years is a crucial part of the process. If consumers are not willing to spend more on carefully made products, it will be difficult for designers to help eliminate the overconsumption that characterizes our time period. The solution to ending overconsumption is convincing people to change their buying patterns; the goal is not spending less or more money than previously, but rather buying fewer (but better) things.

Communication, building an aesthetically sustainable narrative, is therefore the key to making a real difference as a designer. The message about aesthetic sustainability must be communicated in a way that allows users to experience its value. For this reason, I will here explore and discuss how to communicate or market aesthetically sustainable value; the communicative aspect of aesthetic sustainability concerns products as "containers" consisting of time, process, and stories, all of which contribute to increasing their value. Additionally, it is paramount to verbalize or visualize how aesthetically sustainable products are intended to last many years, and that, by virtue of this fact, they will cost more than mass-produced items, which are designed with built-in obsolescence—such items do not age well, and they are meant to be replaced on a regular basis, keeping the wheels of production spinning. The goal of aesthetic sustainability, to a large extent, is to disrupt the conventional means of production.

But how do we communicate that a product is charged with time and aesthetic considerations, justifying its elevated cost compared to a mass-produced product? An important concept in this regard is transparency. The consumer or user must be given insight into the thoughts, processes, and craft behind the product in question. The more transparent the design and production processes, the easier it will be for the consumer to understand the added financial cost compared to products with built-in obsolescence. In other words, the story or the *time* and the craft behind a product can help increase the value of the product enough that consumers will want to pay a higher price for it.

Communicating aesthetically sustainable value primarily relies on connotations concerning time. The consumer's connotative system of reference must be anchored to the object's time of becoming and, potentially, its time of being[3] to allow the consumer to connote that the object possesses a complex expression (due to the process behind it or due to its material composition), making it aesthetically stimulating or pleasurable to behold and use for decades to come.

In this context, visual communication is effective, and a concrete strategy, for example, could be literally to show images of the hands that made the product—pictures of hands in the process of drawing or shaping something. Alternatively, designers or companies could publish on their website sketches, or other forms of process imagery (material experiments, for instance), to give consumers a sense of the product's becoming. Communicating about the process behind or the creation of a product in this way could furthermore include consumers in developing principles or concepts for minimizing waste; end-users could also gainfully participate in creating strategies for reducing a product's wear and tear or for repairing damaged products.

Communicating about or branding a product as being aesthetically sustainable might likewise result in a narrative about how the concept of durability has driven the creation of the product. For example, have durable materials been used, allowing the product to age with beauty? Has the principle of flexibility been incorporated into its function and expression? Or has the intention perhaps rather been to create a neutral, minimal expression that can be used in many different contexts and paired with other objects? If so, show it! Involve the user.

When communicating about aesthetically sustainable value, another strategy to employ is drawing on a product's unique value. Magical things are rare![4] And that which is rare or unique is generally thought to be valuable, as it is irreplaceable to a certain extent. Among other things, unique value can be attained by using diverse patterns or materials—or by combining patterns and materials in different or "random" ways—in a production series. Unique value can also emerge by integrating traces of the production process in the final product or by allowing a certain degree of randomness to influence the design process. Of course, it is possible to draw on unique value in mass-produced objects by imitating artisanship, and, in this way, create an artificial glimmer in the final product. This latter method can be effective, but such an artificial process cannot benefit from the storytelling about the slow genesis of the product, the artisanal techniques, or the *hands* behind it. As a result, the concept loses both transparency and credibility.

As mentioned above, *transparency* is the general rule that all communication about aesthetically sustainable value must adhere to. The user must feel part of whatever lies behind a given product and the intentions that it is charged with. The user should feel enriched with a form of *insider* knowledge about what is behind or what came before the product.

### What is valuable?

In order to speak about value—and aesthetically sustainable value—it is important to relate to what is generally considered valuable according to the time and culture that form one's context. In other words, what is the underlying assumption[5] about value—and, in the same breath, the underlying assumption about the good life—at the time or concerning one's target group. In this regard, it is recommended to begin by analyzing and comparing the target group's culture with the Zeitgeist.

What is considered valuable, of course, depends on whom we ask; however, in the west, stress and ambition lead to a near-universal dilemma: What should we focus our energies on? Where do we place our valuable time? Should we primarily focus on self-realization and career or on family and friends? Of course, one does not rule out the other, but this choice and dilemma permeates everyday life for many. Ought we stay at the professionally enticing seminar at work, or ought we rather pick up the kids early and make hot cocoa and biscuits for them at home? Should we tend to our own personal interests and ambitions, striving for excellence in the workplace, or is it wiser and more rewarding to prioritize family and friends above everything else? Rarely are we able to balance the two in any kind of ideal way, feeling that, in the best Maslowian sense, we have managed to realize ourselves fully as individuals and, at the same time, as parents, lovers, or friends. It is as if the equation simply cannot be balanced. Consequently, everyday life is accompanied by a constant guilty conscience, which can result in a sense of lack: Should we not be at home playing with our children instead of being at an afternoon conference at work (because is it not exactly by spending time with our children that we find the meaning of life)? Or ought we not immerse ourselves in the latest scholarship in our field instead of having coffee with friends? It is because of these kinds of dilemmas that the philosophy of presence, referred to as "mindfulness," has become so popular recently—because presence is a rare commodity in late-modern life. Value—or what is considered attractive—is thus often connected with "rare commodities." For example, time and energy, as well as thoroughness and uniqueness, are valuable elements in a time and culture where most do not have either the time or the energy to commit to thorough and engaged work processes—which is why most things are mass-produced and anonymous. It is for this reason that unique, handmade products—or products that imitate such qualities—appeal to so many late-modern, western consumers. Moreover, this is why focusing on these qualities when communicating about aesthetically sustainable value makes sense.

## From CSR to DSR

CSR, or corporate social responsibility, has to do with the way in which companies include social and environmental concerns in their business strategy. For a company, working with CSR might mean collaborating with producers/suppliers to improve social and environmental conditions for workers and local communities. It might also mean for companies to delineate a set of guidelines for ethical production practices, which might include provisions for guaranteeing that child laborers be provided with schooling by the subcontractors that the companies deal with. Companies seeking to integrate CSR strategies into their business concepts can, then, make a number of demands on their subcontractors regarding human rights, or they may come up with strategies for battling climate change.

DSR, designer social responsibility, or design-oriented social responsibility—a concept I have developed based on my own considerations of and experience with how designers can work with sustainability—is similar to CSR in many ways. The focal point of DSR is an active consideration of human or social circumstances when developing business concepts, but whereas CSR initiatives concern entire, relatively large, companies and strategically must be incorporated into every part of the corporate structure, DSR is considerably more simple and straightforward in scope. DSR is located with individual designers, and the main focus is how designers throughout their work processes can improve human conditions or develop a general practice that will benefit the greatest number of people or small communities possible.

The responsibility for sustainable progress cannot be placed solely with consumers and their willingness to pay more for slow design and ethically produced goods; it is imperative that the industry accepts a sizeable part of the responsibility. As representatives of the industry, designers are able to promote sustainability practices by working with aesthetic sustainability but also with social responsibility— or by combining aesthetics with social involvement and by making decisions about societal and other current issues.

The social responsibility of designers can be expressed by preserving craft traditions; by teaching design processes and thereby passing on their professional expertise; and, finally, by encouraging local production and facilitating the work of artisans. If integrating older, perishing craft traditions as well as creating job opportunities for local communities can be made part of one's design, then one can be said to have integrated DSR into one's work process.

Contributing to the creation of sustainable solutions for a group of people is another way for designers to engage with DSR principles. Choosing local production facilities is one strategy for putting DSR into practice. Another option would be to collaborate with artisans and/or designers from local communities with a decidedly different aesthetic than one's own and, in this way, *remix* or combine the two approaches—thereby developing both forms of expressions.

DSR principles can help increase the value of a product. And if the value is thus increased—meaning that one's goods are sold at a higher price than if they had not been "charged" with social responsibility—it becomes easier to help local communities and commit additional resources to developing one's own processes. It is my experience that an aspiration to profit from sustainability initiatives is often met with suspicion or considered inappropriate by the consumer and fellow designers, since working with sustainable initiatives, ostensibly, should be driven by ideology rather than a desire for personal gain and prestige. However, the fact remains that if one's design is unprofitable, one's business concept is unsustainable. Unprofitability not only affects one's bottom line, it also affects the people one had intended to help in the first place. Similarly, unsustainable business concepts are unlikely to influence patterns of consumption, which might otherwise help minimize overconsumption. Actually, profiting from one's design or business concept ought to be considered as a positive foundation and as a criterion for success, instead of as an expression of economic greed. Without proper funds, it is impossible to produce slow design or pay one's contractors and artisans a fair wage that will allow them to concentrate completely on their work. Furthermore, in the absence of profitability, developing and sustaining traditional crafts or creating design methods for minimal waste production are impossible to achieve unless one has the necessary capital to invest in the time that a given project demands.

Sustainable design products are aesthetically durable, but they are also based on sustainable business concepts—concepts that will actually result in continuous profit.

### *A specific example of DSR*

Recently, I visited Sri Lanka with a colleague and a group of students from the BA program in Sustainable Fashion at the Copenhagen School of Design and Technology. During our visit, we worked on a project about artisanship, cross-cultural communication, and aesthetics, which resulted in a range of clothing pieces and accessories. The project provided me with insight into how designers and design students can integrate long-established local craft traditions—in this case, traditional bobbin lace—into modern design. In this way, visiting designers can contribute to maintaining craft traditions and updating or developing the expression of the specific craft or aesthetic.

I have previously described the importance of integrating consistency as well as variation into aesthetic sustainability.[6] Similarly, when integrating traditional craft techniques into design products, it makes sense to focus on maintaining traditional work processes alongside adding to the original expression a new aesthetic or temporal dimension. Makers working within long-established craft traditions are in danger of creating obstacles for themselves by clinging to a particular aesthetic or expression that was originally tied to the technique itself. To survive, it may be beneficial to combine the original technique with an aesthetic that is not traditionally linked to the specific craft. It may also be worthwhile, in this regard, to embrace the aesthetic qualities of other cultures. As a case in point, the Sri Lankan lacemakers that we interacted with were both highly capable of and very interested in collaborating with visiting designers.

The Sri Lanka project came about as a collaboration between our students and the Dickwella Lace Centre, which is located in the southern province of Sri Lanka in the small town of Dickwella, on the shore of the Indian Ocean. The centre was established following the tsunami of December 2004, and its purpose was to give local women the opportunity to provide for their families and help rebuild the community.

We arrived in Dickwella without knowing exactly what to expect. Right away, we were impressed by the women's kindness and hospitality as well as by the pride they took in their craft and skill. Based on our overly abstract sketches (which had been created without knowing anything about the local craft tradition), they managed to "translate" our designs into concrete patterns and lace, which astutely captured the desired aesthetic, and which could then easily be implemented into the pieces and accessories that the students had initially sketched. The Sri Lankan women then proceeded to "paint" using their lace technique our complex, asymmetrical compositions that differed from the more traditional, symmetrical patterns they had been used to producing (Guldager and Harper 2015). The pride they took in their work and their level of enthusiasm was astonishing. They quite clearly responded to the challenge and made the project part of their own craft.

The Sri Lankan lace-makers were thus highly motivated to challenge their own craft by accommodating a foreign aesthetic. They possessed a fundamental openness and curiosity that inspired us greatly. Their approach to solving the design "riddle" we presented them with in turn inspired them to expand on their own aesthetic in order to accommodate our taste. They very much wanted to "stretch" and further their craft to test its limits.

Integrating traditional bobbin lace—or other traditional, time-consuming crafting techniques—into a design object is a way of creating an aesthetically sustainable narrative involving the time of becoming.[7] The hands and the underlying process thus become visible and shape a story inherent to the traditions and knowledge that have been passed down through generations; at the same time, a tangible bond between the artisan and the end-user is established. The designer's task, responsibility, and challenge are to facilitate the process of integrating the craft elements in such a way as to appeal aesthetically to the end-user as well as to maintain the core of the craft expression. In order to do so, a thorough understanding of the maker, the technique, and the end-user is required.

The will to expand on a traditional craft or aesthetic is key to maintaining and progressing traditional crafts: this involves a combination of consistency of form, which is a product of age-old techniques, and a renewal of the original expression through outside influence. By being open to new aesthetic expressions and the will, as well as the desire, to progress and to incorporate different ways of using their craft, the Sri Lankan lace-makers demonstrated an understanding of this crucial aspect of aesthetic sustainability. The students left Sri Lanka with a desire to continue using lace or other traditional craft techniques in their future work to keep developing their newly acquired design knowledge. Additionally, they experienced what happens, when one, as part of a design process, has to communicate without words (due to a language barrier) and is "forced" to rely on sketches and shape experiments as one's only source of communication. When faced with a language barrier during a design collaboration, the designer must either be very clear when sketching, or else she must be willing to let the collaboration itself shape the final product. The latter is what happened in the Sri Lanka project. Due to our lack of a thorough understanding of the craft (i.e., traditional lace-making), we had come up with abstract and rather unclear sketches that were fairly difficult for the Sri Lankan women to "translate." However, they managed to translate the abstract sketches by using their incredible skill and imagination to turn them into actual laces that could be used as garment details and accessories, whereas the students had to abandon control of the process in turn. The design results were even better than what we had planned or expected. They became actual collaboration pieces instead of mere "products" based on the master plan of a designer. The outcome included design objects that were infused with the hands-on collaboration and the wordless communication that took place that winter in southern Sri Lanka. DSR is precisely about creating the frame for a fruitful design process rather than controlling every step of it.

Traditions are not there to be changed, but intercultural collaboration has the power to connect different techniques and aesthetic expressions. This kind of mutually beneficial exchange helps ensure the progression of design in general and can also help prevent design methods or techniques from stagnating.

DSR initiatives, such as the Sri Lanka project sketched here, can help expose local artisanship to designers unfamiliar with the tradition and vice versa. When a traditional craft expression is compelled to change due to external impulses or *disruptions*, such as our abstract and imprecise sketches, it has a better chance of surviving. Hence, the designer's responsibility is to support traditional crafts by creating fruitful disruptions and finding a balance between consistency (the core of the craft's expression) and variation. The designer should, in other words, provide local artisans with the means to showcase the flexibility of their craft and gain recognition for their work on a wider scale. The point is not to remove the essence of traditional aesthetics from the craft technique associated with it; there is a wealth of stories and beauty connected to each local craft tradition, which must be preserved. The aim, rather, should be to encourage points of confluences between local craft techniques and design strategies in order to promote the inherent flexibility of aesthetics. This approach can help preserve and promote traditional techniques and crafts as well as increase the exposure and interest for these. With an increase in exposure, the value of handcrafted products will automatically rise as well.

A specific aim of the Sri Lanka project was to find a way of simplifying the technique or, rather, minimizing the production time involved in creating the very intricate lace patterns. Instead of creating entire clothing pieces out of lace, the students opted to incorporate lace *details* into their final designs. This had to be done without sacrificing the complexity and detailing that characterize the lace craft. By using lace details instead of working with large areas of lace—which are very time-consuming to produce—makers of the traditional craft will be able to more easily turn a profit on their goods. Another upside is that by relying on detailing rather than creating entire outfits—and by making their patterns more asymmetrical and abstract—their designs will more seamlessly overlap with the kind of minimalism that tends to define the contemporary western "look." This form of translation has the potential to open up new avenues of collaboration and distribution networks that would be of high value to the local producers and their craft.

## Aesthetic nourishment

Aesthetic nourishment is related to the experience of beauty and the feeling of being pulled out of the everyday for just a moment. Aesthetic nourishment happens as a result of being moved by beauty, in an overwhelming sense, in the context of new or familiar surroundings or objects. The previously quoted passage from *My Struggle*—where the narrator is suddenly moved by the mood, the light, and the beauty of a random, everyday scenario—illustrates precisely this kind of experience: ". . . the pleasure that suffused me was so sharp and came with such intensity that it was indistinguishable from pain" (Knausgaard 2012: 307). Experiencing something almost *painfully* beautiful—which has the power to turn an otherwise meaningless, dull train ride or walk into something significant, almost spiritual—is not easy to shake. This kind of experience tends to linger for a good while, as it stores itself in the mind and in the senses.

Aesthetic nourishment is *nourishing* in the sense that aesthetic experiences are "stored up" in the body and in the mind. It is accompanied by a great sense of pleasure that can be described as edifying, as one's subsequent encounters with the world are colored by this experience of beauty.

Beautiful, sublime, or aesthetically nourishing things or phenomena can be "satisfying" in a nearly spiritual way, or in a manner, at least, that cannot be termed purely physical or purely reflexive. Aesthetically nourishing things satisfy us in a way entirely different from food, physical comfort, or products designed to support one's identity and status. Surrounding ourselves with aesthetically nourishing things and placing ourselves in aesthetically nourishing surroundings as often as possible can be hugely beneficial. As Danish Philosopher Ole Thyssen puts it:

> Inhabiting the aesthetic mode, one leaves behind everyday considerations, especially the preoccupation with existence and relevance. Instead, this mode of being opens to a slow, careful, fluid, attentive sensing of the world, which grants access to a host of sensuous options than what was previously available.
>
> (Thyssen 2005: 29; transl.)

However, someone could be more or less disposed, or more or less open, to having aesthetically nourishing experiences. Or someone could be more or less conscious of how important they are.

Designers can help rouse the recipient's senses and mind by creating products that are sensorially stimulating and durable in the way that they have the potential to be experienced, continually, as being interesting, beautiful, challenging, or comfortable "to be around." Aesthetically sustainable products are precisely characterized by, time and again, affording the recipient aesthetic nourishment.

### *The difference between aesthetic sustainability and emotional durability*

> Well this is a knitted cardigan jacket, in wool and it was knitted by my great grandma. Her name was Sigrid as well. She's the one that I got my name from. I'm the fourth generation that's used it. I really like the colours and it's really high quality but I think it's been passed down because we know that she has put so much time into making it.
>
> (Local Wisdom, User Practices)[8]

Project Local Wisdom, discussed in Chapter 3, boasts an abundance of user surveys, the aim of which is to clarify the parameters involved in relation to the kind of clothes that people decide to keep and care for as well as possibly decide to repair and update. The project encompasses several interesting examples of sentimental value and emotional bonds between owners and pieces of clothing; nevertheless, there are examples pointing in a somewhat different, and less emotional, direction.

The quote above describes a girl's attachment to a cardigan passed down for generations that was knitted by her great-grandmother, who, like the girl herself, was named Sigrid. The cardigan is of course charged with many different emotions and stories that help define its particular, subjective value. But, additionally, the cardigan's colors and craft quality are emphasized, which is interesting in relation to the value of aesthetic sustainability. If the cardigan did not nourish the girl aesthetically, or if it did not contain aesthetically sustainable value, I wager that she would not wear it. She would probably appreciate it and store it in her home, and she would probably not want to get rid of it, but she would not wear it. She would definitely not wear it very often. It would not be one of her favorite sweaters, and she would not feel like it was "her." In other words, if the cardigan were merely charged with sentimental or emotional value, it would not have as great a value to her. Precisely the fact that she, time and again, finds aesthetic nourishment in the sensuous encounter with the exquisite wool of the cardigan as well as the satisfying visual experience of its color harmonies (which, by the way, largely correspond to Itten's color theory in relation to, e.g., the contrasts of complementarity and saturation)[9] holds significant meaning. The Pleasure of the Familiar influences the aesthetic experience of the cardigan; moreover, it is charged with several temporalities: the time of existence, as it has a long history going back four generations; the time of becoming, as the girl's grandmother knitted it by hand and spent many painstaking hours creating its elaborate symmetrical pattern; and a brief of time of being, as it is functional, easy to detect and immediately appealing due to its harmonious color combinations.

Aesthetic sustainability is more than the emotional attachment to a product. Of course, there are significant similarities—or, rather, aesthetic sustainability consists *in part* of emotional value. However, whereas the emotional attachment to an object is mainly individual and based on sentimental value, subjective stories and experiences, or memories, aesthetic sustainability is largely universal. Aesthetically sustainable objects are broadly satisfying and they maintain that satisfaction for years, as they evoke universally aesthetic parameters and moods, which either occasion the Pleasure of the Familiar or the Pleasure of the Unfamiliar. Exactly for this reason, it is possible to work strategically with creating aesthetically sustainable products.

In the final chapter, "Aesthetic strategy," I will render concrete the elements of aesthetic sustainability, and, in this regard, I will elaborate on how designers can plan the aesthetically sustainable experience of a product, which, time and again, can nourish the recipient's senses and mind in an aesthetic way.

## Notes

1 Cf. Chapter 5, "The magical thing."
2 Cf. Chapter 4, "Designing the temporal object."
3 Cf. Chapter 4, "Designing the temporal object."
4 Cf. Chapter 5, "The magical thing."
5 Cf. Schein's use of the term "basic underlying assumptions." See the section on "Zeitgeist analysis" in Chapter 3, where I provide an overview of Schein's model, consisting of the artifact level, the value level, and the basic underlying assumptions.

6 Cf. Chapter 3, "The expression of flexible sesthetics."
7 Cf. the section "The time of becoming" in Chapter 4.
8 www.localwisdom.info/use-practices/view/61
9 See the section on "The universal effect of color" in Chapter 1, where I describe Itten's color theory and his seven color contrasts.

# 7    Aesthetic strategy

The experience of beauty cannot be as subjective as it first appears to the person affected by it. If the productive effort to create beauty is to have any meaning at all, then it must be supposed that our experiences of beauty are, at least to a certain extent, shared. [. . .] The artist, the designer, the architect will want to know what he or she has to do to ensure that a public will experience his or her objects or arrangements as beautiful. And to say what the artist has to do would be the task of aesthetics.

(Böhme 2010: 23)

According to the German philosopher Gernot Böhme, different experiences of beauty are similar in kind, despite their apparently subjective nature, and are therefore characterized by a certain universality or by being common to all (or most) human beings. That is probably why we enjoy reading fiction about the experience of unity in the world, or why we get caught up in film sequences about other people's sensuous experiences of beauty. If there were no similarities between different human beings' experiences of the beautiful and the sublime, it would not make sense that we feel a certain satisfaction characterized by the joy of recognition or understanding, compassion and identification, when exposed to a description of other people's enlightening, beautiful or harmonious aesthetic experiences. Precisely this point—that the aesthetic experience is characterized by being universal, and that it thus defies national, cultural and Zeitgeist-based differences—forms the basis of my definition of aesthetic sustainability. Furthermore, this universality is fundamental to the aesthetic strategy that is the focal point in the final chapter of the book.

## Generalization and aestheticization

Like Böhme, Lyotard considers beauty to be universal. What is more, in the essay *After the Sublime*, he makes an important distinction between taste preferences and the pleasurable sensation that beauty can afford the recipient:

If one likes a flower for its colour or a sound for its timbre, this is like preferring one dish to another, a question of idiosyncrasy. This type of empirical pleasure cannot hope to be universally shared. If on the contrary a given singular taste is to be that of anyone and everyone, as is demanded by the pleasure brought about by beauty, this promise can only be grounded on the form of the object procuring that pleasure.

(Lyotard 1991: 138)

As Lyotard points out, taste preferences are often based on individual experiences and memories; in other words, they are altogether subjective and emotional—as in, for example, preferring white hyacinths to blue violets, pear pie to apple pie, or Miles Davis to David Bowie. Because of the subjective nature of taste preferences, they are very hard for designers to control. In contrast to highly subjective preferences, the pleasurable sensation caused by the beautiful (or the sublime) is universal, in the sense that it is what Lyotard describes as "gathering up the manifold" (Lyotard 1991: 138). By this, Lyotard suggests that the experience of beauty contains common elements, elements that are shared universally.

Common elements are worth striving to identify. In order to create an aesthetically sustainable design product, designers must attempt to synthesize the common elements of the beauty experience into the product. The common elements of the beautiful experience make a product experience relevant, extraordinary, and durable.

But how can this be achieved? Rather than preaching the superior beauty of the white hyacinth, one should attempt to pass on the atmosphere and aesthetic nourishment that the white hyacinth encompasses. Everybody has the ability to recognize the atmospheric experience of being sated aesthetically, whether or not they agree that white hyacinths are more beautiful than blue violets.

When singular circumstances or experiences are *generalized* and *aestheticized*, they evolve from being mere taste expressions to become something more. According to Lyotard, it is *the way* that diversity is synthesized—rather than the content or matter itself—which determines whether the expression of beauty is transformed from being merely about individual taste to appealing broadly or to having a more general human character. So rather than integrating a personal experience or memory into a design product—which could for example concern the designer's grandmother and her beautiful, warm home, her carefully made dinners, and slowly knitted sweaters—the story should be "lifted" onto a meta-level and thereby transform into something more general, something that the recipient can relate to, as it connotes "homeliness" and "slow clothing." The personal element of the inspirational source must, in other words, be transformed into something general, so that it can be experienced as relevant and applicable by others than oneself—and in order for it not to close in on itself and thus shut others out. To generate general and aesthetic appeal, moving beyond merely adding personal taste and memories to the design product, designers must transform their subjective experiences of beauty into more general themes or concepts. By converting sensory and mentally strong memories or influences into something more universal or something thematic, it becomes possible to communicate these to the recipient and thus potentially prompt her to have similar experiences. As part of the design process, the individual element must be generalized and aestheticized in order for it to hold both a sufficient degree of relevance and an openness that makes it possible for the recipient to embrace and integrate into her own personal reality and history.

As previously mentioned, in connection with the anecdote about my son's magical treasures,[1] for instance, different experiences of beauty, despite having individual characteristics, are often similar which Böhme and Lyotard also point out. Such similarities can be defined and implemented into objects, and thus made useful in the design process. Correspondingly, Aristotle points out in *Poetics*, in relation to the cathartic experience (Aristotle 1996), that a performing artist (or designer) should focus on universal human themes in order to "move" the recipient. Pursuant to catharsis, which is associated with the theater experience and more specifically with tragedy rather than comedy, this will typically result in themes such as "separation", "unrequited love", or "death". The cathartic experience implies namely that the viewer, from a safe place (the theater seat), is overwhelmed by the feelings associated with loss, but is able to process them without being in any actual danger.[2] One can, however, only feel moved or affected if the play (or accordingly the design product) is perceived as relevant and applicable to one's own life. And the experience of relevance and applicability will only occur if the author of the tragedy (or the designer) manages to generalize and aestheticize the story (or the product) to such an extent that it has broad appeal. In order for the recipient to be moved or feel affected, and thus paving the way for the aesthetic experience to occur, the author (or the designer) must make use of universal human themes that can synthesize the manifold sources of inspiration as well as generalize and aestheticize individual taste preferences and memories. Generalization causes "openness": due to its stylized form, the story/product may accommodate or contain the recipient's own needs and feelings, thus rendering the experience inclusive.

An excellent example of generalization and aestheticization is Karl-Ove Knausgaard's novel *My Struggle, Book One* (2012), which I have previously quoted. Despite the story's highly individual character and the author-narrator's presence in the text, the novel has a wide appeal, as Knausgaard's bases his otherwise very personal story on universal human themes such as "the father figure," "mother fixation," "sibling rivalry," "loss," "self-esteem or lack thereof," "depression," and "family-career frustrations." Most people can relate to and identify with such themes—especially the postmodern Western human being who deals with a multitude of roles as parent, partner/spouse, friend, professional, sibling, neighbor, and colleague on a daily basis. Hence, the novel is relevant and "open;" the trivial and the manifold are generalized and aestheticized.

Implementing aesthetic value into a design product involves thematizing influences and individual experiences in order to simplify and aestheticize them, thereby creating relevant and "open" objects or concepts. The designer should seek common denominators of beauty experiences and engage in experiments that can lead to ways of passing these on through an object or concept. This could mean implementing tactile experiences or color harmonies, as well as symbolic linguistic messages, into objects.

Working with generalization and aestheticization implies, in other words, the transformation of singular experiences into something generally applicable, or the transformation of subjective experiences into moods that can be experienced as relevant across and despite time, place, and context. This transformation largely concerns giving the singular, individual starting point of the design process an understandable applicable *form* or expression. In the novel *Slowness*, Milan Kundera explains the process: "Imposing form on a period of time is what beauty demands, but so does memory. For what is formless cannot be grasped, or committed to memory" (Kundera 1996: 38). If one is unable to transform the essence of the mood that one wishes to pass on to the recipient into *form*—and thus aestheticize and transform singular, particular memories and influences into something general to be disclosed, articulated, or expressed in shapes and materials—the product/ concept will be experienced as fleeting and irrelevant.

## The durable expression

In the previous chapters, I have attempted to answer the question of how best to create a long-lasting design product, a design product that could continuously provide the recipient with aesthetic nourishment, thus making it aesthetically sustainable. Over the course of this book, I have detailed different approaches to working with aesthetic sustainability, in terms of product or concept design. In the following, I will summarize and fortify these methods.

The underlying assumption of this book is that an object can easily be made durable, both in terms of quality and functionality, without being perceived as aesthetically appealing by the recipient. But if the aesthetic appeal is not present— if, say, the object is simply not perceived as beautiful or fascinating or in accordance with what the user wants it to express—it will not be cherished or kept for very long. In that case, the object cannot be described as aesthetically sustainable.

In the context of the universal experience of beauty, the last sentences of this chapter's epigraph are worth highlighting: "The artist, the designer, the architect will want to know what he or she has to do to ensure that a public will experience his or her objects or arrangements as beautiful. And to say what the artist has to do would be the task of aesthetics" (Böhme 2010: 23). My ambition is exactly this: to clarify what the designer needs to consider in order to ensure that the recipient of a design product has an aesthetic experience. That is why I have created links between philosophical aesthetics and the design experience.

The aesthetic strategy to be unfolded in the following sections is both a collection and an operationalization of the book's reflections and case studies; at the same time, the aesthetic strategy opens up towards further work on the topic. The strategy is designed for practical use, and it therefore contains concrete guidelines for working designers that can be used in the creation process of a product, whether tangible or intangible. In addition, the strategy can be used to analyze design objects or concepts.

The strategy is designed based on the hypothesis that people need aesthetic experiences, beautiful experiences, and aesthetic nourishment; additionally, everybody sometimes needs to be challenged, fundamentally, and sometimes needs to be confirmed in her basic assumptions about the world. The strategy aims to enable designers to work intentionally and strategically with the creation of an aesthetic user experience that can strengthen the bond between product and user.

The idea for developing the aesthetic strategy originally arose from the experience of lacking a concrete and applicable "hands-on" link between philosophical aesthetics and design; I needed a "tool" and a vocabulary when working with design aesthetics, in particular between aesthetics and the design process (i.e., not as the starting point for design analysis). The ideological thrust behind the strategy is to make aesthetic considerations an active part of the design process. The result is a strategic tool that enables designers to integrate considerations about the aesthetic design experience into the design process as well as to articulate the aesthetic intentions behind a product.

If designers are looking to create "market potential" by designing products that match the consumer's need to "belong" or by trying to boost the recipient's ego, the products in question will most likely have a very short shelf life. If one, on the other hand, strives to create products that are meant to meet and satisfy spiritual and aesthetic needs, rather than solely functional and socially fleeting desires, these are likely to "speak to" what Walker has described as our highest human potential (Walker 2007: 50). By doing so, the untenable or fleeting can potentially be overcome. A product of such a caliber can only be described as durable and sustainable.

## The contradistinctions of aesthetic strategy

> The artist is the hand which plays, touching one key or another, to cause vibrations in the soul. It is evident therefore that colour harmony must rest only on a corresponding vibration in the human soul; and this is one of the guiding principles of the inner need.
>
> (Kandinsky 2008: 62)

Working with the aesthetic strategy tool involves a strategic planning to make the recipient's soul "vibrate," in Kandinsky's understanding. When designers use the aesthetic strategy, they must thus include the intended recipient's experience as a part of the design process.

The strategy is structured around four conceptual pairs that constitute each other's opposite. The pairs are inspired by the distinction between the beautiful and the sublime. Throughout the book, I have been using several of the concepts. The four conceptual pairs are as follows:

- Instant payoff versus instant presence
- Comfort booster versus breaking the comfort zone
- Pattern booster versus pattern breaker
- Blending in versus standing out

Designers wanting to give the recipient an aesthetic experience in line with either instant payoff or instant presence should primarily seek to appeal to the recipient's

# AESTHETIC STRATEGY MODEL

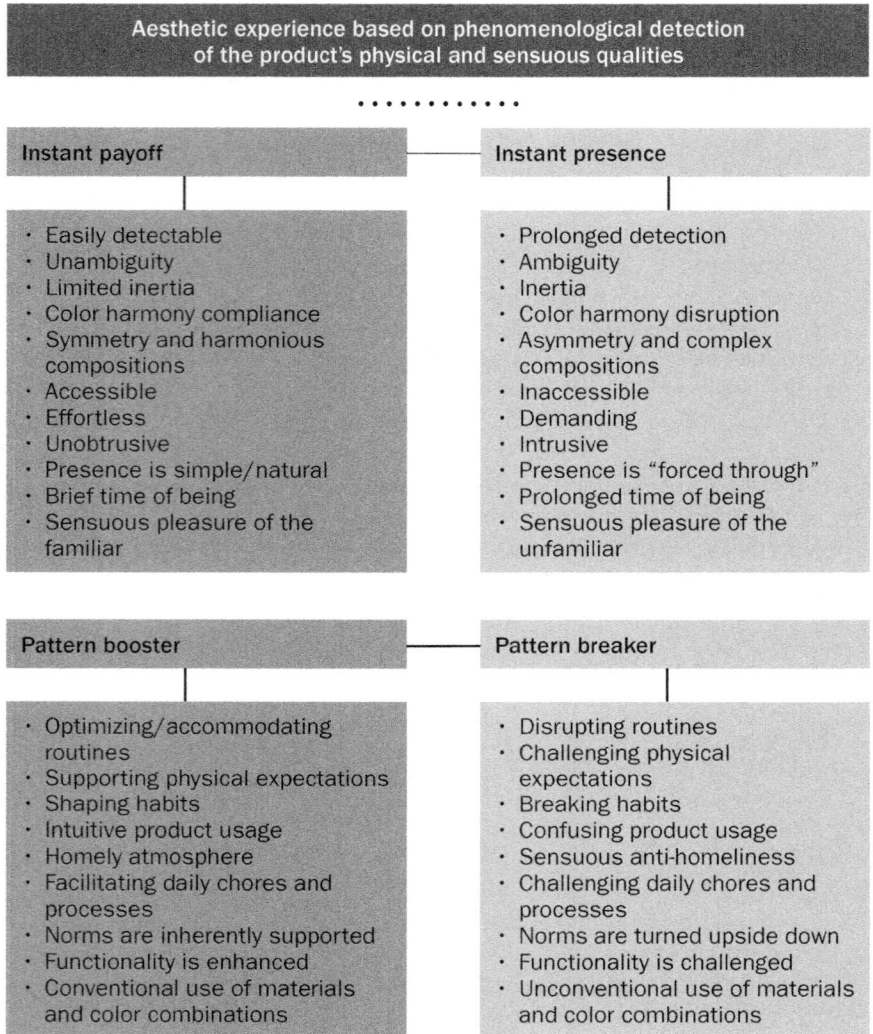

> **Aesthetic experience based on phenomenological detection of the product's physical and sensuous qualities**

· · · · · · · · · · · ·

### Instant payoff ———— Instant presence

| Instant payoff | Instant presence |
| --- | --- |
| · Easily detectable<br>· Unambiguity<br>· Limited inertia<br>· Color harmony compliance<br>· Symmetry and harmonious compositions<br>· Accessible<br>· Effortless<br>· Unobtrusive<br>· Presence is simple/natural<br>· Brief time of being<br>· Sensuous pleasure of the familiar | · Prolonged detection<br>· Ambiguity<br>· Inertia<br>· Color harmony disruption<br>· Asymmetry and complex compositions<br>· Inaccessible<br>· Demanding<br>· Intrusive<br>· Presence is "forced through"<br>· Prolonged time of being<br>· Sensuous pleasure of the unfamiliar |

### Pattern booster ———— Pattern breaker

| Pattern booster | Pattern breaker |
| --- | --- |
| · Optimizing/accommodating routines<br>· Supporting physical expectations<br>· Shaping habits<br>· Intuitive product usage<br>· Homely atmosphere<br>· Facilitating daily chores and processes<br>· Norms are inherently supported<br>· Functionality is enhanced<br>· Conventional use of materials and color combinations | · Disrupting routines<br>· Challenging physical expectations<br>· Breaking habits<br>· Confusing product usage<br>· Sensuous anti-homeliness<br>· Challenging daily chores and processes<br>· Norms are turned upside down<br>· Functionality is challenged<br>· Unconventional use of materials and color combinations |

## Aesthetic experience based on semiotic decoding of added symbolic value

. . . . . . . . . . . . .

### Comfort booster

- Basic assumptions are supported
- Short decoding time
- Easy to understand
- No surprise
- Comfort
- Connotations anchored in linguistic messages
- Homeliness
- Comforting
- Unselfconsciousness
- The Beautiful
- Emotional pleasure of the familiar (the comfort zone is padded)

### Breaking the comfort zone

- Basic assumptions are challenged
- Prolonged decoding time
- Hard to understand
- Surprise
- "Something happens!"
- Uncertain connotations
- Uncanny discomfort
- Cathartic
- Forced self-awareness
- The Sublime (retention in the chaotic "phase 2" of the sublime experience)
- Reflective pleasure of the unfamiliar (the mind is expanded)

### Blending in

- Encapsulation of contemporary beauty
- Following the crowd
- Speaking the "language" of the Zeitgeist
- Clear signals; rapid decoding
- Being a part of something/ someone
- Cohesion
- Easily decoded product-signals and value
- Homogenous/uniform
- Simplicity
- Anonymity
- Camouflage
- Balance
- Tradition
- Neutral expression

### Standing out

- Intentionally going against the flow
- Breaking patterns
- Sampling and a mishmash of signs
- Prolonged decoding
- Standing apart from something/ someone
- Confused signs and symbols
- Challenging product-signals and latent value
- Heterogeneous/non-uniform
- Complexity
- Individuality/attention-getting
- Eye catcher
- Imbalance
- Novelty
- Inventiveness

senses and bodily presence. If the emphasis is on either boosting or breaking the recipient's comfort zone, designers should rather focus on working symbolic value or aesthetically sustainable value into the product. The same goes for the two other conceptual pairs. Working with pattern boosting or pattern breaking is related to the phenomenological, sensuous aesthetic experience, whereas blending in and standing out are more closely associated with either meeting or challenging the recipient's basic assumptions.

Generally, it should be noted that all the "dark grey" categories in the following strategy model are linked to the Pleasure of the Familiar, while the "light grey" categories are characterized by the Pleasure of the Unfamiliar.

Thorough knowledge of the target audience is a prerequisite for working with the aesthetic strategy; a product can only be experienced as *aesthetically nourishing* if it is perceived as relevant by the recipient. The first step of working with the aesthetic strategy should therefore involve conducting an in-depth audience analysis (which advantageously can be structured using the three steps of Schein's model: the artifact level, the espoused values, and the basic underlying assumptions [Schein 2004: 26]). The second step in the strategic process should clarify whether the product should meet or challenge the recipient's basic assumptions or expectations. In respect to this stage, consideration of what makes sense in relation to the product category one is working within is appropriate. Finally, as the third step of the aesthetic strategy, one must clarify which elements of the strategy one wishes to work with, and naturally *why* this would make sense in relation to one's design aim. This step should also consider how the strategy process should proceed: which idiom and connotations will support the established aesthetic strategy?

The following sections will thoroughly describe the content of the as well as the scope of the contradistinctions of the conceptual pairs. In addition, I will clarify how to include the strategy in the design process and how to conduct a design analysis using the critical terms of the model.

However, as a starting point, the model on pages 134–135 illustrates the conceptual pairs and their differences.

## Instant payoff versus instant presence

Seeking to provide the recipient of a design with an experience of either instant payoff or instant presence implies that the recipient is either able to instantly detect and use the design object, or that she is being challenged due to the materials, shapes, and colors used, making her suddenly feel very present or torn from her daily "hypnotic" activities. The category of instant payoff/presence is especially useful in the creation of physical objects, as opposed to intangible design concepts or experience design, since the focus is on sensuous experiences. In the instant payoff/presence category, the recipient's sensuous handling of the object is the center of attention, and the designer should therefore aim at making the recipient *detect* rather than *decode* the given design object.[3] The focal point of this category is the object experience itself rather than the connotations triggered by the object.

No matter which type of aesthetic experience one aims to provide the recipient with—or in what way one strives to nourish the recipient aesthetically[4]—thorough knowledge of the target audience is essential. In relation to instant payoff/presence, it is particularly important to obtain knowledge about the recipient's bodily or sensual basic assumptions. For example, what does the recipient expect when introduced to sitting room furniture? Are there certain materials she specifically associates with chairs, benches, or sofas? Are there certain colors, color combinations, or patterns that are generally linked to the product category in question? Only by knowing what the recipient expects is it possible either to meet or to challenge those expectations, and thus to provide the recipient either with a sense-based Pleasure of the Familiar or with the opposite.

### *Instant payoff: accommodating tactile expectations*

Designers working with an aesthetic strategy including instant payoff should basically meet the recipient's sensuous expectations, meaning the recipient's anti-cipations regarding the sensations of the product—how it "should" feel touching, holding, lifting, or using it.

A particularly satisfactory instant payoff experience can be triggered by imbuing a product with an immanent or obvious way of using it, thus rendering it transparent. This would imply that the product itself, without the use of supplementary words in the form of hang tags or other such text, can "explain" to the recipient how to use it. Consequently, good packaging design is usually characterized by instant payoff; users are able to detect intuitively or after a short tactile investigation how to open, operate, and close the packaged object again. This kind of product experience is typically highly satisfying. The same is true for many other product categories that advantageously can be distinguished by emphasizing how functional and simple they are to detect and use. As part of the instant payoff experience, the aesthetic quality of an object must be easy to detect as well as how to use the object immediately. This aesthetic category should therefore lead to creating unobtrusive and accessible objects of harmonious expression.

The aesthetic experience of instant payoff is straightforward and characterized by an instant connection between the object and the recipient. The recipient gets what she expects, or maybe an even more "tailored" product experience than she could ever imagine. However, it is important to point out that one should not imitate or copy existing products in order to accommodate the recipient's tactile expectations. Rather, the design practice should include experimenting with an idiom that can provide the recipient with the pleasurable feeling of everything being as expected and simultaneously remaining intriguing for an extended period of time (and thus also aesthetically sustainable). As previously described,[5] the aesthetically sustain-able object *juggles* with both familiarity (that resembles something the recipient is already familiar with and hence has seen or touched before) and rejuvenating variation. The rejuvenating or regenerative element can, in relation to instant payoff, be relatively understated and for example involve the use of materials that can easily adapt to the shape of the object, and thus fit the idiom perfectly, but at the same time differ slightly from materials that would usually be used in similar objects. Maybe they work even better. Incorporating an instant payoff strategy into the design process can advantageously include a study of how much renewal an object, such as a coat, a chair or a coffee cup, can hold: How far can you stretch the shape or the sensory qualities that characterize the object, and still be confident that it can give the recipient a pleasurable instant payoff experience?

Working with instant payoff as one of the building blocks of one's aesthetic strategy includes having the following guidelines in mind:

- The product should basically be quickly and easily detected; the usage of it should ideally even be a part of the idiom. The instructions for use, so to speak, must be inherent.
- Functionality should be in focus, akin "the most beautiful spoon is the spoon which is best at being a spoon and hence is made of a material that can actually withstand being used for cooking and eating."[6] The object should be immediately usable and functional and "talk" to the user's hands.
- The recipient's physical and sensual expectations must be met; if the design object appears heavy, it should be heavy, and if the object looks soft, it should be soft.
- Non-complex, symmetrical, harmonious structures are typically perceived as easily detectable, and should therefore dominate the product expression in accordance with instant payoff.
- The material must "fit" the shape—or it should have a minimal degree of inertia in relation to the object's shape.[7]
- The experience of the object should be characterized by a short time of being.[8] It must, so to speak, be easily accessible.

All the contradistinctions that make up the aesthetic strategy should be regarded as spectra on which the designer can place her product, depending on whether she wants to meet or challenge the recipient's expectations. If one works with an aesthetic product strategy that involves challenging the recipient without bringing her completely out of her comfort zone, the placement on the instant payoff versus instant presence spectrum could appear as outlined below in Figure 3.

Instant payoff

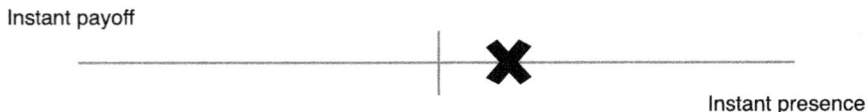

Instant presence

*Figure 3* Instant payoff versus instant presence

The product is clearly placed within the instant presence category, but close enough to the instant payoff category that the aesthetic experience may include some of the elements that characterize this category.

This location leads me to the opposition, or the prerequisite, of the instant payoff experience: instant presence. All the contradistinctions of aesthetic strategy are prerequisite in the sense that it generally doesn't make sense to work towards the creation of a challenging aesthetic experience, if one hasn't initially clarified what a non-challenging aesthetic experience consists of. And vice versa.

## Instant presence: challenging sensuous assumptions

"The more different temporalities are deposited in an object, the richer and more complex its appearance" (Ørskov 1999, 84; transl.) As shown by this quote, working different time courses into an object will make it appear complex and hard to detect, according to Ørskov—and when working with the instant presence category, complexity is desirable. Hence, if one *charges* an object with both the time of becoming and the time of existence and simultaneously (or due to this) aims for a prolonging of the time of being,[9] the object-accessibility will be remarkably reduced, and thereby the recipient might experience the Pleasurable of the Unfamiliar.

A product that is charged with different temporalities or that appears complex due to, for example, asymmetry or the use of a material that doesn't seem to match its shape—and which is thereby *inert*—is intrusive. It forces the recipient to stop. It forces her to be present. And the sudden presence is the core of this aesthetic category. Aesthetic nourishment in an instant presence way includes being forced to be present, forced to relate to the object one is facing or holding, and forced, more or less brutally, out of the daily grind. An aesthetic instant presence experience is often characterized by self-awareness; the recipient's expectations are undercut, and she is thereby confronted with her experiential limitations.

> The arts, whatever their materials, pressed forward by the aesthetics of the sublime in search of intense effects, can and must give up the imitation of models that are merely beautiful and try out surprising, strange, shocking combinations.
>
> (Lyotard 1991: 100)

As the quote points out, according to the French philosopher Lyotard, surprising, strange or shocking combinations (of shapes, materials and colors) can be effective agents if the designer is aiming to pull the recipient out of her daily hypnotic chores, or "force" her to be sensuously, physically present. The fact that something (unexpected) is happening can cause the pleasurable instant presence experience.

If the designer works with instant presence as a jigsaw piece of her aesthetic strategy, it will generally include aiming for an extension of the time of being or the detection time—and thus retaining the recipient in *chaos* or in the second phase of the aesthetic experience[10] for an extended moment with the intention of inducing the Pleasure of the Unfamiliar. In order to achieve this, one can work with one or more of the following guidelines as a part of the design process:

- Disrupt universal aesthetic principles for composition and harmony by using, for example, asymmetry or interfering with color harmonies. Working with such disruptions is a way of challenging the visual sense, since the eye of the recipient is unable to immediately find balance and structure, and to capture or conceptualize the object. Thus, the time of being is prolonged.

- Repeal gravity in the sense that the designer can "play" with something that looks light, but is in fact heavy. For example, one could create an illusion of lightness by imitating lace or other featherweight fabrics as print on a heavy material—or simply create a hole-pattern in a compact material and thereby visually break its dense surface.
- Redesign focusing on using materials and/or objects in new ways or in new contexts. An example of this could be finding inspiration in the principles of ready-made art and surrealism[11] and placing objects that belong in one particular context into another, for example, creating objects for outdoor use that imitate indoor objects, or experimenting with "surprising, strange, shocking combinations" (Lyotard 1991: 100) as surrealist artist Meret Oppenheim did with her fur-lined teacup in 1936.
- Incorporate visual tactility, or the illusion of texture in a medium that normally doesn't provide the recipient with tactile experiences, for example, visual media such as websites and smartphone apps. It may be effective in relation to the work of instant presence to incorporate the illusion of texture into a visual medium, and thereby "force" the eyes to imagine how it would feel to stroke the presented surface—and thus *feel with the eyes*. By linking the senses in this way, they are strained.
- Charge the object with time. By underlining the time of becoming or the design process *in* the object itself, whilst emphasizing the time of existence, the complexity of the object will be increased, and the time of being will be prolonged. For example, by creating an illusion of wear or decay, several different time courses are deposited in the object.

## Comfort booster versus breaking the comfort zone

The dichotomy of comfort booster versus breaking the comfort zone covers partly emotional pleasure and partly a challenging extension of the recipient's consciousness. If the designer chooses to implement elements from this category into her aesthetic strategy, she should strive to awaken emotions and connotations within the recipient—and to either rapidly anchor these or intentionally let them "float" and remain elusive, at least for some time. Semiotic decoding is, in other words, the basis for both the aesthetic experience that can boost the recipient's comfort zone and the experience that can burst it. Comfort booster versus breaking the comfort zone is attached to the recipient's mental activity and connotations, and the added value attached to objects and phenomena. Comfort booster versus breaking the comfort zone can accommodate both material and immaterial design. This implies that one, in contrast to the previously described phenomenological-based contradistinction of instant payoff and instant presence, can certainly use elements from these categories in the development of a concept, an event, or a service design—and in addition, of course, in the creation of product design.

Basically, comfort booster is attached to the Pleasure of the Familiar and the beautiful aesthetic experience, since the comfort supporting aesthetic experience contains an immediate pleasure, which is characterized by accessibility. Breaking the comfort zone, on the other hand, is closely related to the Pleasure of the Unfamiliar and the sublime aesthetic experience—particularly the postmodern sublime, as Lyotard outlines it in "The Sublime and the Avant-Garde," in which he characterizes the sublime as a heady experience of *"Something happening"* or suspended privation (Lyotard 1991). The aesthetic, comfort zone breaking experience includes thus, in line with the other "light grey" categories of the aesthetic strategy,[12] the delightful feeling of something (unintelligible) happening before one's eyes. This delight contains a hint of fear, since the consciousness is not instantaneously able to capture and understand the object or phenomenon that initiates the experience.

In the following two sections, I will elaborate further on the comfort boosting and the comfort zone breaking aesthetic experience, and provide guidance for integrating these aesthetic categories into the design process.

### Comfort booster: padding the recipient's comfort zone

If the designer chooses to include the comfort booster category as one of the building blocks of her aesthetic strategy, she should basically aim for "padding" the recipient's comfort zone. The recipient must feel at home and secure, and this security feeling should hold the pleasant sensation of knowing what is expected of you and of being able to meet those expectations. As previously pointed out, thorough recipient knowledge is also required in this category; if one doesn't know the recipient's expectations, these are very difficult to meet. When working on adding a comfort boosting aesthetic experience to a product, one of the pillars is to create a sense of complicity between the recipient and the product. The recipient should feel "seen" or "heard"; the connotations that arise in the encounter with the product should convey emotions that are comfortable and familiar. Having a comfort boosting aesthetic experience should feel a bit like having a cup of hot cocoa served, accompanied by a good confidential talk with someone you love when you most need it. You feel seen and heard—and you feel like you are in good hands. Products that pad one's comfort zone have a similar ability to "embrace" you and provide you with what you need, when you need it, as well as to make you feel comfortable and "at home."

Comfort boosting design products focus on semiotic decoding, and, as previously described, everything within the world of semiotics is viewed as signs.[13] This implies that all objects contain meaning and messages that go beyond their purely physical existence or their form, color, and materials. Thus all concepts are *charged* with narratives and with values that can only be decoded correctly if you know the cultural codes that under e them. The messages or narratives therefore only have an effect on the recipient if they make sense to her, or if she is familiar with the "code language" they are embedded in. In relation to the comfort booster category this is particularly important; the designer must become thoroughly acquainted with whoever she is targeting and aiming to reach and please. What do these people believe in? What are they influenced by? What do they find interesting, beautiful, or cool? Most importantly, how can one meet and support these basic assumptions? If the recipient considers the good life one involving slowness and time for reflection, the designer must meet this need and charge the product or concept with easily decodable time of becoming or with linguistic messages that clearly indicate *slow living*. A concept that would support a need like that could be a café that would explicitly advertise slow food and "space for contemplation."

In the comfort booster category as little as possible should be left to interpretation. There must be a high degree of clarity; the added values should be immediately decodable. The recipient must, in other words, be filled with the satisfactory feeling of "what you see is what you get." Therefore, in relation to the comfort boosting aesthetic experience, transparency is a relevant concept. This is not in a physical sense, as it applies in the instant payoff category, in which transparency must be understood in the sense that the product instructions should be intrinsic or inherent and "readable" by the recipient's hand. Transparency in relation to the comfort boosting aesthetic experience should be based on the recipient's thoughts and imagination—and be expressed in clear signals, clear messages and easily understandable instructions, linguistic as well as visual. The recipient must, more or less immediately, be able to decode the product's added value. Hence, there should not be any confusing signals or ambiguous messages in the comfort boosting aesthetic design experience. The connotations that the recipient experiences need to be rapidly anchored, and should not be allowed to "float"; the decoding time must be short. The recipient should not be in doubt about what the product can do, will do, and signals. An effective way of anchoring the recipient's connotations is by using linguistic messages, which could be a product title, description on a tag, an accompanying brochure, or other supplementary words. The designer can also choose to work with visual anchoring by assembling various visual elements, each of which can anchor each other's meaning.[14]

In order to be able to anchor the recipient's connotations in the most appropriate way, the design process must contain a clarification of the recipient's connotative frame. The connotative frame is a person's frame of reference, which states all of her embedded beliefs or interpretations. It is largely conditioned by the social exchange of views and cultural ties, and thus formed on the basis of experience. When one has clarified all the possible connotations that could arise within the recipient's mind when confronted with the design product or concept, one must identify which of these connotations are appropriate or desired, and which are to be regarded as totally inadequate. The recipient's inappropriate connotations should be eliminated, while the applicable ones must be fixed or anchored—which, in relation to the comfort boosting aesthetic experience, means anchored to the sphere of familiarity, comfort, and security. The recipient should experience the joyful sensation that characterizes the Pleasure of the Familiar.

To summarize what characterizes the comfort booster category, and to also add a few extra dimensions and concrete examples in order to illustrate the characteristics of the category, I will conclude this section by presenting some guidelines designers can use when working with padding the recipient's comfort zone:

- Aim to make the recipient feel safe and at home by creating products that arouse connotations of home ness. Integrating such connotations into one's aesthetic strategy thorough recipient understanding is crucial. To this end, the designer must obtain a basic understanding of what triggers the recipient's feeling of homeliness. Is it awakened by the use of fonts that carry connotations of handwriting, or photos that imitate polaroids or make use of analogue references? Or is the recipient's feeling of being "at home in the world" rather associated with being on the road and in motion, for example, with suitcases, airports, and takeaway products? Having insight into the connotative frame of the receiver type one is targeting makes it possible to incorporate, for example, homely references, which hold the possibility of the creation of an immediate bond between the recipient and the product.
- Incorporate transparency by letting the product be easily decodable and understandable. This could, for instance, come from the integration and support of prevalent myths about "the good life".
- Integrate aesthetically sustainable value into the product by working with social sustainability, akin to the Sri Lanka project,[15] and incorporate storytelling about this as a value-creating factor. Or one could create a number of sustainable dogma[16]—such as local production, the use of recycled or organic materials, or directions on minimal wash—and make them a part of the brand narrative. Knowing that the products they buy and use are created with the intention of protecting the environment and/or to improve human conditions may be experienced as comfort boosting by a wide range of consumers.

If the designer wishes to make use of the comfort booster category as one of the building blocks of her aesthetic strategy, she should generally seek to eliminate any ambiguity; the product must be easily decodable and easily usable, and the stories and values it is charged with should be easily accessible and easily understood.

### Breaking the comfort zone: designing unpredictability

It takes time to decode an object or a concept that is characterized by ambiguity or complex signals. And this time lapse can be painfully pleasurable! The comfort zone breaking aesthetic experience is one of being awed by an object or concept. Or it can be an experience characterized by being pulled, more or less brutally, out of one's comfort zone, because of a sudden confrontation with an object or a concept that is at once fascinating and on the verge of being repulsive. As opposed to the comfort boosting aesthetic experience that supports the recipient's basic assumptions, the backbone of the comfort zone breaking experience is disturbing and questioning basic assumptions. Or perhaps even transforming them.

As shown in the aesthetic strategy model, the aesthetic experience that is associated with the disruption of one's comfort zone bears similarities to the cathartic experience. As part of the cathartic experience, which Aristotle discusses in *Poetics* (Aristotle 1996),[17] the experiencing individual undergoes emotions that can be described as unpleasant—loss, unrequited love, jealousy, anger, frustration, etc.—but in an aestheticized form. For example, when watching a theater performance, the viewer is presented with the characters' dilemmas and anguish and becomes "wrapped up" in the beauty and thematization of the story. Because these negative experiences are presented to the recipient in an "digestible" form—even though they will still move her and exaggerate her feelings and emotions—she can absorb them, process them, and thus "safely" endure them. Similarly, the design product or concept that manages to give the recipient a comfort zone breaking experience might force the mind and imagination into working overtime and exaggerate accompanied emotions. Thus, an aesthetic experience that borders on the uncomfortable, but holds its own particular pleasure, follows. This kind of aesthetic experience is, due to its dialectical movement between the pleasurable and unpleasurable, potentially groundbreaking and broadening.

When the designer works to prolong the design product's decoding time, which characterizes both of the challenging "light grey" categories based on the decoding (breaking the comfort zone ands standing out),[18] her objective should be that the aesthetic product experience leaves a lasting impression. This impression is typically based on a chaos experience. However, it is crucial that the comfort zone breaking experience also contains understanding or an "ok, now I get it" experience. The recipient must, so to speak, be thrown into the deep end and experience the chaos that follows—but the chaos must cease again. The duration of chaos experience can be more or less extended, but it is crucial in order for the overall experience to be aesthetically pleasurable that it does eventually cease. Therefore it is important to once again emphasize the importance of thorough recipient knowledge when defining an aesthetic strategy as a part of the design process. One needs an understanding of how far one can take the chaos experience, or perhaps more correctly, how far it is necessary to take it in order to move the concerned recipient. It may take a lot of chaos to momentarily throw the recipient off course if the recipient type one is targeting is accustomed to seeing a large variety of things on a daily basis; or it might just require a few vague signals and ambiguous messages to affect the respectively more sheltered recipient.

If the designer uses breaking the comfort zone as one of the building blocks of her aesthetic strategy, she can integrate one or more of the following elements:

- The unexpected and the unpredictable—or the experience of "something happening" (Lyotard 1991: 100)—which in this context could result in surprise elements, such as obscure or ambiguous linguistic messages; for example, a product title that doesn't directly match the product, but rather intentionally creates confusion, and thereby prolongs the decoding time.
- "Playing" with the recipient's connotations. If one, as a result of initial target group studies, has found out that the targeted segment typically associates status and luxury with "traditional" status symbols, such as jewelry made of noble metals and Scandinavian furniture classics, one can turn these connotations upside down by deconstructing a furniture classic and recycling the materials into new products (and using the story of this process as corrsponding storytelling), or by creating jewelry from recycled materials.

- A hodgepodge of fragmented stories revolving around the moods one wishes to share with one's recipient, and which can arouse almost surreal connotations. This could lead to a collage of narrative fragments, visual and verbal, with no direct link, which could seem like interrupted glimpses into dreams. A collage of stories will typically cause a prolonged decoding time; the recipient is kept in doubt for a prolonged moment that makes it very satisfying to finally reach an understanding.
- An integration of themes that are considered taboo in the culture that the target group is part of, or that are not usually dealt with in design products. These could include deformity, physical disabilities, obesity, old age, or decay. Working on aestheticizing and conceptualizing such themes involves forcing the recipient to face them. One might even incorporate the selected theme in an understated way or "hide" it, so that the recipient will experience the shock effect of suddenly, after a prolonged decoding time, understanding the concept. By "fooling" the recipient into being captivated by the expression and the shapes, colors, and material combinations, only to then to realize that the theme of the immediately appealing object is quite surprising and different than expected, the dialectical movement that also characterizes the catharsis experience will occur. The recipient is at once attracted and repelled. Such an experience has the potential of being highly comfort zone busting and horizon broadening. Having presumptions about what you find attractive and repulsive is usually greatly limiting. Perhaps the deformed, crooked, aged, wrinkled, and decayed hold a special beauty that is overlooked by Western cultures due to their predominant fascination with youth. Perhaps the overlooked beauty of the crooked contains a great deal of aesthetically sustainable value.

Generally, working with breaking or blowing up the recipient's comfort zone means extending the chaotic phase, or phase 2 of the sublime aesthetic experience.[19] However, it is important to emphasize that the overall category of comfort booster versus breaking the comfort zone should be viewed as a spectrum on which one, as part of one's design strategy, can place the design product/concept. The designer can choose a position close to the center, thereby including elements from both sides. Yet, as mentioned in the previous section on instant payoff versus instant present, making a design strategic choice is important—which means that even though the product/concept position is close to the middle of the spectrum, one must choose to work with either a predominantly comfort supportive aesthetic experience or a predominantly comfort breaking experience. In Figure 4 the chaotic phase is extended to a maximum; from this strategic point, the recipient must be blown away and held in a state of uncertainty and chaos for as long as possible.

Comfort booster

Breaking the comfort zone

*Figure 4* Comfort booster versus breaking the comfort zone

## Pattern booster versus pattern breaker

If one chooses to work with elements from either the pattern booster or the pattern breaker category as a part of one's aesthetic strategy, striving either to accommodate and support or to challenge and question the habits of the recipient is the focus. In relation to this aesthetic contradistinction, it is important to point out that the habits that are to be either supported or challenged are the *physical* habits of the recipient, such as routines that are associated with human bodily interaction with and sensuous response to objects or spaces. Pattern booster versus pattern breaker deals with people's routine activities and all the objects that are a part of our daily life. Furthermore, these two categories focus on our daily whereabouts and movements in both public and private spaces. The categories are, in other words, particularly useful in relation to product and spatial design.

The universal aesthetic principles, which have previously been described,[20] are largely based on the fact that human beings, despite differences, are physiologically pretty similar; we have two arms, two legs, roughly the same adult height, ears, nose, mouth, eyes, etc., and our senses work in much the same way. Therefore, it might be possible, despite our value-based and culturally based differences, to establish guidelines on how shapes, colors, and materials are perceived and "consumed" by the senses, as well as on what is considered the most balanced and immediately understandable expression *a priori* (before interpretation and added value). As an example, some colors seem immediately warm (red-purple-orange shades), while we experience others as cool (blue-green hues).[21] Likewise, some shapes and compositions are experienced as "quiet," harmonious, or balanced, while others appear dynamic, discordant, or out of balance.[22] Certain objects will guide viewers to a particular way of using them or to certain patterns of behavior, simply due to their shape; the instructions are, so to speak, immanent or inherent and thus wordless. Other objects will require detailed instructions before they can be used. A pattern boosting object will thus, due to its design, support the most natural use of it—and will in addition typically make use of color harmonies and shapes that are experienced as immediately appealing by the human sensory system. In contrast to this, an object that can be described as pattern breaking will intentionally confuse the user and prolong the time it takes for her to find out how to use it.

In the same vein, some urban spaces and other public and private spaces encourage habitual movements and uniform behavior, whilst other spaces intentionally challenge habits and routines of their users. Some spaces provide comfort and relaxation with soft shapes and warm colors, while others dictate movement and efficiency by making use of cool colors and streamlined forms.

Similarly, an object can contain an inherent direction of intended usage. As a simple example, some seating furniture will invite you to sit down, relax, and let *slowness* take over, while other pieces will indicate that this room is one where you shouldn't stay long (which can be useful in waiting rooms and transit halls). Likewise, lamps and light can invite specific behavior; while subdued and warm lighting generally signals "peace and togetherness", strong, clear light usually appeals to efficiency and "getting things done."

Objects and spaces can—on the basis of their shape, colors and/or materials—support specific behavior and habits, just as they can be evocative. Hence, their "instructions" can be inherent. The designer can integrate physical and sensory qualities in objects and spaces, and invite certain behavior or actions (pattern booster), or the designer can consciously seek to challenge, and perhaps change, the user's routine behavior (pattern breaker).

### Pattern booster: nursing the creature of habit

The pattern boosting aesthetic experience should provide the recipient with a sense of having daily rituals and habits supported, or perhaps even optimized. The habitual usage of objects and daily surroundings are in focus in this category, and the recipient's physical expectations are supported and accommodated. Pattern boosting objects propose almost "hypnotic" behavior; they must support the user's habits, rhythms, and daily activities. The design should therefore be adapted to everyday routines and accommodate the user's most natural interaction with the physical world and its objects.

As part of the pattern boosting design experience, routines and daily activities are supported. Creating products that fit the recipient's daily space and rhythms is therefore an obvious way to work with integrating the pattern booster category in one's design strategy. Spaces, both public and private, which are designed and furnished in line with the pattern booster category, should thus support the most natural way of moving through and in them. Similarly, objects that are designed to be pattern supporting must relate to the simplest usage and seek to promote natural interaction. The pattern supporting design experience should generally feel like someone placing a gentle hand on your shoulder, showing you the way and helping you through your daily chores in a way that is understated and comfortable.

The designer can advantageously, when working with pattern booster, make use of conventional materials, or materials that "match" the shape she is "molding." This means that if one is creating a sofa, one should upholster it with conventional sofa materials, such as wool, canvas or leather; if one is designing clothing, one must strive to meet the recipient's tactile expectation that clothing is generally soft and wraps itself around the contours of the body, and hence not engage in "stiff" experiments by integrating recycled plastic or wooden slats. The material must, within the pattern booster category, be characterized by having minimal inertia in relation to the shape it should take.[23] It should be experienced as obvious in relation to the given object, and it must meet the recipient's expectations of which materials are typically used within the given product category, and thus be within the recipient's frame of experience or connotative frame.

Intuitive usage is, besides pattern support and minimal inertia, an important pattern-boosting element. A good example of this is a tablet for reading e-books which imitates the ordinary physical book. As a result of this imitation, the product is easy to use; it enables the user to intuitively—by hand—interact with the object. Flipping pages, as with a traditional, physical book, is straightforward to grasp. Thus, when a digital medium, such as a tablet, creates an illusion of traditional physical book pages made of paper, it "speaks" to the recipient's hands rather than to her head, thereby creating a sensuous connection between the object and the subject. Furthermore, this example contains an interesting coupling between "the best of both worlds," and thus an optimization of habits and patterns: There's nothing better than browsing through a traditional book, and only a few are willing to completely abandon the sensory experience of reading a physical book made of paper in favor of the e-book. The worst thing about traditional books is that they take up so much space—especially when one is traveling and wants to bring a lot of good reading material along. On a tablet, one can download thousands of books, and even organize them in what resembles physical bookshelves.

This way of working with product design can easily be transferred to other product categories. It is important that the product supports the most natural, physical, and sensory interaction and usage; meets the expectations of the user; and makes life easier for her. The illusion of tactility is, in the above tablet example, not intended to confuse the recipient and intentionally prolong the decoding time (as is the case in the comfort zone breaking aesthetic experience), but rather to create an intuitive, immediate connection between the product and the recipient. On top of that, the appeal to touch and explore works as an inherent user manual.

Like the other contradistinctions of the aesthetic strategy, pattern booster versus pattern breaker is to be considered a spectrum on which one chooses a position in accordance with one's design strategy. Figure 5 illustrates a pattern boosting design experience located in between pattern booster and the transition to pattern breaker. Thus it contains some rejuvenation. In working with pattern booster as an aesthetic category, one should not—in accordance with the instant payoff category—merely copy existing products in order to accommodate the recipient's need for pattern boosting products. The practice should be to experiment with an idiom that can provide the recipient with the pleasurable experience of having her routine-based needs met, whilst containing the potential to remain interesting and aesthetically pleasing. One must, in other words, aim for adding expressive or aesthetic value to the product.

Pattern booster

Pattern breaker

*Figure 5* Pattern booster versus pattern breaker

If the designer wishes to use pattern booster in her aesthetic strategy jigsaw puzzle, this can be done by, for example:

- Supporting or even optimizing the daily chores and the daily whereabouts of her recipient. The need for pattern supporting aesthetic experiences are highly situational; we all have routine tasks that we conduct every day, and that are preferably easily done. When working with pattern booster, the designer should demonstrate that she understands her recipient's situational needs by striving to make the individual's life as comfortable and easy as possible. This can take the form of products that are functional and easily usable, and which are simultaneously pleasant to look at and touch.

- Allowing the object to appear familiar and recognizable by only making minimal changes that can optimize and extend the object's appearance-based lifespan or aesthetic sustainability. A cup should basically look and feel like a cup. A coat should look like a traditional coat and thus meet the recipient's expectations of what a coat looks like, but it can be optimized aesthetically by beautifying details or the use of harmonious color and material combinations.
- Referring to familiar objects that the recipient cares for, and possibly *sampling* the best of two product categories in order to make life easier for the user, while nourishing her aesthetically, as in the tablet example.
- Incorporating inherent instructions. The simplest example of an object that through its design or shape shows how it should be used is a pair of finger gloves; the five fingers *say* "Put your fingers here." Similarly, a space can "tell" the user what to do, or how she should move through it, simply by letting the shape of the space or the objects and furniture within it indicate specific usage or lead her on her way.
- Embracing habits, as well as allowing habit formation. There is a high degree of satisfaction in experiencing your habits being accommodated. If you, for example, love drinking hot tea with both hands wrapped around the cup, a cup without a handle will support this habit, and thus be easily understandable, despite the fact that it slightly challenges the shape of a traditional cup.
- Supporting the most natural activities in public spaces, so that hospital visits, municipality meetings, or visits to the zoo can be carried out with ease and experienced as pleasant. This could be done through the design of guidance or signs. Generally it should be easy and straightforward to find your way in public spaces; if not, it will likely be a source of frustration. The pattern booster category is linked to physical aesthetic experiences through an integration of physical direction givers that naturally lead the way and eliminate confusion. These can be an effective way to create pattern support, and to gently help the recipient on her way. Examples of physical direction givers could be bumpy flooring that can guides the recipient's feet in the right direction, or weak or strong lighting that can indicate whether or not one is moving in the right direction. There is a large variety of alternative ways to lead people in the right direction. Traditional signs typically make use of words and pictograms. The advantage of using alternative, more sensory methods is that one avoids cultural confusions caused by the recipient not understanding the language or not being able to decode the illustrations correctly. Pattern booster is one of the categories in the aesthetic strategy that should generally set the stage for the creation of products that are universally easy to read.

Generally the designer, when working with pattern booster as a part of her aesthetic strategy, should boost and even optimize the recipient's habits and routine behavior.

The recipient must feel that the product or the space "knows" her so well that it "understands" what she needs and what she finds pleasurable, and supports her daily routines.

### Pattern breaker: waking the creature of habit

Striving to provide one's recipient with a pattern breaking aesthetic experience implies stirring her up and challenging her habits and routines—or rather questioning why she does things the way she does. Perhaps one purposely makes the product "difficult" to use, or perhaps one "forces" the recipient to take a break in her daily, routine-based activities by making use of unconventional materials or surprising details. Through its expression, the pattern-breaking product should question, transform, and/or break habits and routine-based behavior. Hereby, the pattern breaker category is associated with the Pleasure of the Unfamiliar. As previously described, the Pleasure of the Unfamiliar is a "troublesome" kind of aesthetic pleasure, in the sense that it isn't immediately accessible or comfortable.[24] It challenges and questions the familiar, and forcefully extends the mind of the recipient, but precisely therein lies its strength.

One of the most pattern breaking experiences I can think of is the experience of having to navigate in a country in which one has never been before, and which has a fundamentally different culture than one's country of origin. The mismatch of different letters (that makes signage unhelpful), unfamiliar sounds, and smells, as well as a multitude of very strange objects and products, might for the first days of one's journey be quite overwhelming. When you can't even find the most basic items in a supermarket, or when you don't know how to use the shower in the hotel room because it doesn't work the way showers "normally" do, you are forced to stop and deal consciously with every little thing you are doing. Nothing can be done automatically. The pattern breaking experience is, in other words, anything but a sense of homeliness and comfort. Pattern breaker is all about anti-homeliness. But pattern breaker also holds the exceptional pleasure that comes along with anti-homeliness: although it is inconvenient, annoying and, most of all, challenging not being able to find one's way and not being able to use even the most ordinary objects, it is also exciting and pleasurable to be forced to cope with challenging situations akin to the supermarket and shower situations (which are, however, not exactly life-threatening). The pattern breaker category implies a degree of chaos—but not unceasing, overwhelming chaos.

But how does a designer charge a design product with anti-homeliness or the physical experience of being completely out of one's depth? Pattern breaker relates, as mentioned and in line with the pattern booster category, to the recipient's sensory experience of the tangible world, which is why the physical, sensory interaction with (everyday) objects or the recipient's whereabouts in (everyday) spaces are in focus. If one is to add anti-homeliness to a product and appeal to the recipient's senses, this should be done by challenging the recipient's bodily and sensory expectations to the tangible world as well as disrupting her patterns or routines. One should strive to question whatever the recipient takes for granted. To do this, a dominant part of the design process should be turning upside down the sensory basic assumptions associated with the product category one is working with. This could involve anything from interfering with the way people usually sit in a chair or by a table, or the way coats are worn, to a critical assessment of what a coat even is, or how a table's sensory qualities can provide the user with continuous aesthetic nourishment. The purpose of the pattern breaker category is to give the recipient an aesthetic experience that holds sensory Pleasure of the Unfamiliar.

Breaking a pattern may involve challenging the way the industry does things. This could be made manifest in the creation of an opposition to the fashion industry's collection cycle represented by SS (spring-summer) and AW (autumn-winter) collections. Creating aesthetically sustainable clothing is a way of challenging the fashion industry's artificially made need to constantly buy new stuff, and replace old garments even if they are not at all worn out. By creating clothing with multi-functional elements, which imply that they can be used in spring, summer, autumn and winter, or by designing aesthetic and quality durable clothing, one is taking steps towards breaking the predominant collection cycle, while challenging the recipient's consumption patterns.[25]

Thus, being a pattern breaking designer or working with pattern breaking elements may include departing from the predominant use-and-throw-away culture. Designers can "force" recipients away from the fast moving pace of mainstream consumption by encouraging them to take care of their things and repair them if necessary. If one is, for example, working with infusing traditional craftsmanship and design products, as we did in the Sri Lanka project,[26] one is adding sensual and intellectual aesthetic value to the product. This kind of added value can "push" the consumers to take better care of the product. Establishing an emotional and aesthetic bond between a product and the recipient can thus be a way of working with pattern breaker as a part of one's aesthetic strategy.

In line with this, a clothing designer can inspire the recipient to change her consumer habits by appealing to the idea of minimal clothes washing or even completely refraining from washing certain garments. This can be pursued by working with materials that neither can nor should be washed, or by creating clothing items that do not "assume their character"[27] until they have been used for a while, or that feel more intriguing to wear after days or weeks of intensive use.

If the designer works with pattern breaker as part of her aesthetic strategy, she should generally seek to provide the recipient with a pleasurable "this is not usual" experience. This can be done in one or more of the following ways:

• Striving to create a glimpse of human beings as the "creatures of habit" we are, and possibly even parodying the human fondness for things being and looking like they always have. This could be done by intentionally complicating daily routines and creating concrete, physical obstacles. By doing this, one can "force" the recipient to stop and question the benefit and advantages of the habitual behavior, and perhaps even change it.

- Challenging the recipient's physical habits by rejuvenating the shape or the material of familiar, everyday objects—and thus waking the user up from her repetitive daily procedures to consider whether or not they are beneficial.
- Designing "imperfect" objects, in accordance with wabi-sabi aesthetics.[28] This objective holds an inherent challenge to conventional beauty. In doing this, the designer can contribute to questioning current consumption patterns. Through object design one can introduce the recipient to a fundamentally different view of what beauty is and, as part of this, challenge the common perception of decay.
- Breaking with cultural patterns and habits such as the use-and-throw-away culture or the tendency to wash clothes excessively, which can be done by creating garments that shouldn't be washed very often, or that should not be washed at all. By this, I don't mean designing clothes that need to be dry-cleaned or hand washed gently, but clothes that become more and more beautiful and interesting through frequent or daily use (perhaps because usage "finishes" them), and that thus create an intimate bond with the user. Such garments become *one* with the user, and assume their character through wear-and-tear.

In general, the pattern breaker category involves habits and routines being challenged and perhaps even transformed. The recipient will, on the basis of the pattern breaking design experience, move on in the world with a new perspective on the way she usually behaves—and perhaps with a desire to change some of her habits and introduce new, more interesting routines, or more durable ways of doing things. Pattern breaker means creating new opportunities through design.

## Blending in versus standing out

Blending in versus standing out is a pair of opposites based on decoding—not only on the recipient's/user's decoding of the product, but also on the decoding of her network or people with whom she shares opinions. These aesthetic categories revolve around the human sense of identity; around whom the subject wants to be; and, around whom she wants to distinguish herself sharply from. Blending in versus standing out draws on the human desire and need to signal identity and moods to fellow humans—whether this involves an urge to "camouflage" and blend in or to stand out from the crowd. The aesthetic value connected to having a sense of identity is the focal point.

Working with the creation of products that are meant to provide the recipient/ user with the feeling that she either fits in or stands out only makes sense if it is first defined exactly *what* the recipient should fit into or stand out from. An analysis of the recipient's context, including the Zeitgeist, must therefore form the basis for the incorporation of blending in or standing out into one's aesthetic strategy.

### *Blending in: wrapping the self in comfortable camouflage*

"Beauty has a community-forming function. In the pleasure taken in certain things, people with the same taste feel united" (Böhme 2010: 27). I have chosen to start this section with this quote from Böhme, since it largely characterizes the pleasurable aesthetic experience of fitting in or blending in. There is a great deal of comfort in experiencing that one, due to one's preferences, belongs to a group of people—and that like-minded people thereby exist. In line with the Böhme quotation, this sensation contains a high degree of community feeling.

Despite the fact that I, as part of my elucidation of aesthetic sustainability, strive to break with "taste" and personal preferences, in order to create general human criteria for assessments of whether products are aesthetically pleasing and long lasting, it is difficult to completely bypass taste preferences and trends that affect the personal taste. But perhaps this is an important point. Perhaps one should not seek to completely eliminate or invalidate taste preferences. As previously mentioned,[29] according to the French poet and philosopher Baudelaire, the beautiful consists partly of an "eternal, invariable element," and partly of a "relative, circumstantial element," which is subject to external circumstance (Baudelaire 1964: 3). A piece of art should, according to Baudelaire, contain an element of eternity or endurance, but it must also be marked by the world's volatile instability in order for it to be experienced as relevant and significant by the recipient. This point can advantageously be transferred to the design object. The object—which is perceived as appealing, attractive, interesting, beautiful, and relevant—may contain, in accordance with Baudelaire's argument, something lasting as well as something fleeting. Or it may consist of both an element that is independent of and impervious to time and place and that remains remarkable and relevant over the years, and an element that is appealing due to its volatility and by virtue of its novelty.

In connection with the blending in category, the volatile element—or the object-element that is attached to the common look of the time, and thus indicates the most prevalent taste preferences—should be relatively prominent. This means that the recipient will blend in with the Zeitgeist-look by acquiring the object.

All categories of the aesthetic strategy have aesthetic sustainability as the focal point—and are thus conceived as a variety of ways that the designer can work towards creating a product that, due to its aesthetic qualities, possesses longevity. For this reason it is important to emphasize that when working with blending in, one should strive to create an expression that can provide the recipient with aesthetic nourishment for years, despite the fact that this category draws on the common taste preferences of the time, or indicates an encapsulation of contemporary beauty. Hence, one should strive to create an aesthetically sustainable core, which in relation to the blending in category would mean complying with universal aesthetic principles,[30] since this category basically sets the stage for very little "noise" or few challenges. At the same time, one should draw on contemporary and habitus-related taste preferences. Thus, working with the blending in category involves a dialectic movement between durability and volatility.

The vast majority of people need confirmation. This is probably why people with similar tastes cling together on social media, assuring each other, for example, that minimalist interior design, spiced up with accessories in simple, graphic patterns and with a few strong colors, indicate good style, or that a "unique" mixture of recycled finds and new products is the right way to decorate your home.

This need for recognition and acknowledgment is largely met by products in the blending in category.

As mentioned, a Zeitgeist analysis is crucial when working with both blending in and standing out. A Zeitgeist analysis can, as previously described,[31] provide insights into the basic underlying, current assumptions as well as the artifacts that reflect them. Using Schein's cultural analysis model as a template for the Zeitgeist analysis is recommended. When working through Schein's model, one must initially, on the artifact level, register objects or artifacts that the target segment considers attractive, beautiful, and/or visible signs of status. As the next step, one must seek an understanding of the segment's espoused values, and finally conduct an analysis in order to reach the basic assumptions that determine the registered preferences. These preferences should thus basically be accommodated.

Considering blending in as a kind of camouflage, the look of an object created in accordance with this category can still be outré and colorful (literally as well as figuratively), depending on the surroundings—whereas a sparrow blends into the Danish countryside, a colorful parrot appears camouflaged in a rainforest. Blending in means, in other words, to fit in, or to merge with the surroundings. So the look of the blending in category is not necessarily neutral and minimalist, which the name of the category potentially connotes.

When working with blending in as a building block in one's aesthetic strategy, one basically aims to meet the recipient's need to appear camouflaged and to not attract attention. Most people know the feeling of sometimes wanting to not stand out, to just fit in or wrap up in comfortable camouflage. Some people have a greater need to blend in than others. The need to blend in can thus be situational, but it can also be linked to a specific recipient type who feels pleasure by following the "middle-of-the-road" approach.

To reinforce the above, here are some guidelines for working with the blending in category:

- Aim to establish an interaction between the permanent and the transient by, for example, creating a product core that respects the universal aesthetic principles—uses color harmonies or a symmetrical, balanced design language etc.—as well as a product part that is changeable, and which therefore may be more trend based. This could, for instance, result in modular garments with a durable core and a number of replaceable parts, which continuously or seasonally can be updated, and thereby follow the "look of the time." It could also result in a piece of furniture that appears neutral and thus can be used in many contexts or easily blend in, but which also has changeable elements that follow the prevailing style of the time, and thereby allow the user to reap the approving gazes from "taste-companions." Or it could be oversized clothing that, due to quality-based durability (a durable core), can be used for a number of years, and that can easily be adjusted according to whether fashion dictates a defined waist and distinct feminine shapes, or loose robes and androgynous references.

- Work with an understated and discreetly elegant expression. Blending in products are not "noisy;" they often offer the user a subtle and flattering look. This could be garments with a flattering fit that highlights the user's figure beautifully, and thereby provides her with the feeling of comfort and confidence that follows from feeling well dressed—in a discreet way and without attracting undue attention.
- Create a fusion between different product elements. An example might be a product characterized by what Itten would describe as the contrast of saturation (Itten 1997: 282–98). The contrast of saturation is characterized by being based on one single color that is "broken" with either white or black, resulting in a tone-on-tone color scale. By working with this color contrast, the recipient can experience that distinct elements of the product unite or flow into each other. Thus, this is a way of creating a merge within the design object rather than a merge between the object and the surroundings.
- Use camouflage as an important part of the aesthetic blending in experience. It can feel pleasurable to be camouflaged, merging with one's surroundings and becoming part of a group of like-minded people or of a beautiful milieu with which one would like to identify. The designer can work with camouflage in several different ways. In the launch or the styling of a product, a fusion between the product and the surroundings can be created in order to support the product's understated elegance and define its affiliation or frame of reference. Camouflage can also be incorporated in a way that more closely relates to the recipient's need to belong and be seen as "spot-on" in relation to the norm or the conforming look. Camouflaging is therefore in this context generally about belonging and cohesion.

### Standing out: basking in attention

If the designer works with an aesthetic strategy with the standing out category as one of the building blocks, she should basically aim to break with social conventions or with whatever is considered usual and ordinary, and thus go against the flow. Standing out is an eye-catching, challenging, and original aesthetic category.

The recipient type who is intrigued by standing out products often has an urge to withdraw from the community, and to display this withdrawal. Both the blending in and standing out category are connected to the recipient's decoding of the product, and thus to the signal value the product is charged with, as well as to the response of those surrounding the recipient to the product. The focal point of this pair of opposites is hereby the recipient's sense of identity and her need to express herself or to signal what she stands for.

In relation to the standing out experience, the need for recognition and affirmation is dominant, though in a rather different way than the confirmation the individual seeks by blending in or by demonstrating her desire and willingness to be "current" due to her consumer choices. The recipient type who aims for standing out from the crowd strives for recognition due to originality and boldness. To her, it is prestigious to stand out.

To stand out may involve the use of strong colors and unconventional color combinations in contexts and situations where others dress in black, gray, subdued, or well-balanced colors. Or it can involve the use of solely black and white in a context where this is not customary. In other words, what it means to stand out from the crowd and challenge the "ordinary," depends entirely on the context. For this reason, it is crucial that the designer gains insight into her recipient's demographic and its whereabouts, milieu, or connotative frame.

To differentiate yourself from the crowd, or from the contemporary, often involves challenging the "look" that is generally is regarded as the most appealing, and thus, it may involve "the aesthetics of the hideous." If you dare to be "ugly" by, for example, drawing on decay, wear-and-tear, distortion, and unconventional combinations of colors, patterns, and materials in contexts or cultures where people generally seek to appear as "neat" and attractive as possible, then you will undoubtedly challenge a multitude of conventions and hence signal a certain distancing from society's norms.

Despite the point made regarding the importance of the context, the standing out category indicates something *overdressed*—literally and figuratively. This category holds an immanent urge to exaggerate and challenge. Even though you can easily stand out by expressing yourself in a minimalist and neutral way in a milieu where everybody else is decorated and flashy, the aesthetic standing out category connotes from its title the outré, exaggerated, wild, and daring. One should therefore, if implementing standing out in one's aesthetic strategy, seek to appeal to the recipient's need to bask in the attention and to be noticed and admired for the courage to be different.

One could advantageously associate standing out with the terms "anti-trend" or "countertrend"—the need to be in opposition to whatever is customary in one's social and cultural context . The standing out experience contains the pleasure of going against the "norm" and sometimes doing this for its own sake. There is a degree of the provocateur within the recipient type who deliberately seeks to demonstrate her disapproval of the "standard" or the "norm" by significantly standing out.

In line with the other three aesthetic pairs of oppositions, the span of blending in and standing out should be considered a spectrum. In Figure 6, the product is placed within the standing out category, but not very far from the transition to blending in. This could imply that the product clearly opposes social conventions by challenging, for example, what it means to be well-dressed, which typically includes the "freshly ironed" and "immaculate." The product could thus imitate

Blending in

Standing out

*Figure 6* Blending in versus standing out

decay or encourage frequent use (apropos the wabi-sabi aesthetics.)[32] At the same time, the product could form a counterpoint to predominant trends by, for example, drawing on androgynous elements if fashion "dictates" a clear distinction between femininity and masculinity. The location of the X on the scale on Figure 6 (fairly close to the middle) also indicates that the expression of the product should hold a certain degree of familiarity and harmony. Hence, by wearing or using the product the user can "flash" her affiliation with a niche segment, or can perhaps be described as a first mover within a rising consumer tendency.

If the designer incorporates the standing out category in her aesthetic strategy, she can draw on one or more of the following elements:

- The unique, and the cultivation of one-off effects, which could be reflected in a visible time of becoming.[33] Emphasizing the design process and "hands" behind the product can make it appear irreplaceable, unique and outstanding. The incorporation of a clear time of becoming in the product also typically prolongs the recipient's decoding time. Furthermore, the decoding time of the people in the recipient's milieu is prolonged, which is desirable in this category, since it is linked to the recipient's need to stand out from the crowd. If the product is decoded as complex, innovative, or unique, the user feels pleasurably different by virtue of her product selection. A product that is visibly marked by the process of its creation will particularly stand out in a context or environment where products usually should appear "finished" and polished.
- Signal bewilderment as a result of a hotchpotch of signals, which can be obtained through a *sampling* of signs from several different "realms" or by a fusion of elements that are normally separate. One can also work with signal bewilderment in the breaking the comfort zone category, but the focal point in relation to standing out is generally the reaction of those surrounding the recipient to the product, rather than the recipient/user's own reaction. The product must of course be perceived as intriguing, unusual, outstanding, or challenging by the recipient herself, and the recipient's connotations, when confronted with the product, should be complex and multifaceted. It is important that the recipient's social circle also responds to the product signals—and preferably reacts with a whiff of outrage or at least with surprise.
- The breaking with social conventions and unwritten rules for how to behave and for which artifacts should be used, when, and for what. This principle can result in the creation of products that confront reputable or prescribed behavior, for example by signaling relaxation and casual elegance in a social context where one is expected to appear effective and "streamlined." Alternatively, the outcome could be products that should be used in a way that doesn't fit their idiom, such as a garment that resembles pants or leggings but is to be used as a scarf, or a chair that is impossible to sit in, and therefore rather than being furniture must instead be considered as a sculpture.
- A clear distinction or definition of sub-elements within the product. To achieve such an effect, it is advantageous to make use of the contrast of extension (Itten 1997: 299–315) and thereby work with expanding the area of color that

visually "fills the most." This can generate a natural eye-catcher; the eyes will seek the area that contains the most visually "heavy" color, and if this color is allowed to be predominant in a larger area than what Itten recommends when aiming for a harmonious composition, a compositional disruption will occur. The eyes of the recipient will be "forced" to stop when they reach the dominant color. In relation to the standing out category, it is not the product's purely physical and sensuous qualities that are in focus, but rather whatever the product symbolizes or the connotations that are attached to it. Therefore, it may be useful to highlight the details of the product that particularly qualify it as rejuvenating or controversial, by means of a rupture of the harmonic quantity contrast. An example could be highlighting the aforementioned integration of the time of becoming by underlining the tracks from the process behind the product using a visually "heavy" color. If one is working with a rupture of social conventions, one could emphasize the details in which this rupture is particularly evident visually. Such visualizations can be effective because, in relation to the standing out category, it is crucial that the outside world relates and reacts to the innovative and provocative elements of the product. After all, it is more interesting to stand out if it is noticed.

- A challenge to what beauty is by drawing on the deviant and quirky. As a part of this, a breakage of the universal aesthetic principles can occur, and thus the eye and the imagination can be "forced" into overdrive mode.[34] However, it could also result in promoting decay and wear-and-tear in a context or milieu in which that is not common practice, such as the aforementioned "aesthetics of the hideous."

## Concluding the work on aesthetic strategy

It is not the intention that when one works with the aesthetic strategy one must make use of all the categories, nor all the pairs of opposites, at once. It is extremely rare that it makes sense to integrate all four pairs of opposites into the design process simultaneously. One should carefully choose the aesthetic categories that are most useful when considering the product category as well as the recipient segment that one is dealing with. This could, for instance, lead to a tailor-made strategy containing the instant payoff and the comfort booster category which, due to these two aesthetic building blocks, sets the stage for the creation of a product that meets the recipient's basic assumptions. A product such as this is typically given in a homey, comfortable mood, which can be supported by symmetry and color and material combinations that correspond well, and that are thereby easily detectable and decodable.

Or you could choose to form a strategy containing building blocks from the categories of instant presence, pattern breaker and standing out—and thus strive to create an aesthetic product experience characterized partly by a complex idiom, partly by functionalities that challenge the way similar products are used, and partly by an unconventional combination of signs and signals.

Or you could establish an aesthetic strategy containing breaking the comfort zone and standing out, and hence focus on the recipient's connotative experience by working with ambiguous linguistic messages and challenging social conventions. When working on intangible design, such as concept or experience design, it makes particularly good sense to focus solely on the creation of symbolic value.

As a designer, it will make sense to concentrate on either a "dark grey" or a "light grey" strategy,[35] and thereby aim to either satisfy and meet the recipient's expectations (dark grey strategy) or challenge these (light grey strategy). One can nevertheless also work with a combination of a dark grey and light grey strategy— even without ending up with an incomprehensible mixture. However, if implementing elements from both the dark grey or expectation-supportive categories and the light grey or the challenging categories, it is important that this is not done to avoid making strategic choices. If the implementation of both dark grey and light grey elements in the product creation is due to an aversion to making a choice or an inability to select the right categories, the product might turn out to be unappealing and irrelevant.

When working with the aesthetic strategy, it is essential to make choices and dare to form a strategy that can create a "sharp" product that is perceived as relevant, applicable and intriguing by the recipient. If the product is not experienced as relevant, it will not be aesthetically nourishing.

## Design analysis with a focus on the value of sustainable aesthetics

The aesthetic strategy model is created with the intention of considering the recipient's aesthetic experience as a significant part of the design process—and thus to be used by the designer. However, the strategy model can also easily be used for design analysis. When applying the model to design analysis, one should "walk through" similar steps as done in a Schein analysis;[36] one must first observe and register, then examine, and finally analyze the collected data (Schein 2004: 25–37). Consequently, in order to use the aesthetic strategy model to conduct a design analysis, you turn the model "upside down." The design analysis will thus review the following activity points:

1   Start by looking at and describing materials, shapes and colors, as well as registering the connotations the product awakens.
2   Consider the values of the designer that may have conditioned these choices; is the designer, for example, working with predominant tendencies, or does the product hold a vision to create a durable look?
3   Determine whether the intention behind the product is primarily to provide the user with a sensory experience or to create symbolic value as the focal point.
4   Analyze the basic assumption behind what can be seen, heard, felt, or read. Is the intention of the product or concept to break with the recipient's comfort zone and challenge her basic assumptions? (And why?) What are the underlying intentions? And, not least, how well are these intentions expressed in the product?

An important part of a design analysis related to aesthetic sustainability is a clarification of the design product's aesthetic, sustainable value—including an assessment of the expression's durability. Such an assessment involves an analysis of whether the product can provide the recipient with aesthetic nourishment. Does the product, in other words, possess qualities that will enable the recipient to feel continuous pleasure by watching it, touching it, and using it—and therefore be inclined to take care of it and repair it, if necessary? Accordingly, one can conduct an analysis to clarify whether the product makes use of some of the guidelines from the model of aesthetic strategy. For example, is the product appealing to sensuous presence by incorporating varied textures that can provide the recipient with a stimulating tactile experience? Or is it striving to optimize everyday behavior and routines by "guiding" the user? A design analysis that focuses on sustainability should point out the product's expressive durability and aesthetic qualities.

The next step in the development of the concept of aesthetic sustainability is to elaborate on the design analytical work—and as part of this, to create an analytical model similar to the model of aesthetic strategy that can be used for both object and concept analysis. This could serve as a tool that companies and designers use when clarifying the extent of aesthetic sustainability in their present design strategies, as well as how more aesthetically sustainable initiatives can be implemented.

The implementation of sustainable aesthetic value in products is a crucial step in the showdown with mindless consumption and overconsumption. If more aesthetically sustainable products are created, a real counterpoint to the use-and-throw-away culture is established.

## Notes

1  See Chapter 5, "The magical thing."
2  See the section on the sublime in Chapter 2 for an extensive description of the cathartic experience. In this section, catharsis is linked to the sublime aesthetic experience that similarly is connected to unique pleasurable pain.
3  For more about the difference between detection and decoding, see the section on "The easily decodable" in Chapter 1.
4  See the section "Aesthetic nourishment" in Chapter 6.
5  See the section on "Aesthetic decay, slow aesthetics" in Chapter 3.
6  See my description and analysis of Plato's interconnection between the beautiful and the good in "The beautiful" in Chapter 1.
7  Inertia is a term used by Willy Ørskov. Inertia concerns how "manageable" or pliable a material is in relation to the shape it must take in order to form an object. If a material is inert, it is difficult to manipulate—i.e., it must be treated or processed before it can be pressed into shape. On the other hand, a material of minimal inertia can easily be made to fit the object's shape. See the section on "The experience of minimal inertia" in Chapter 1 for a further description of the concept of inertia.
8  See the section on "The time of being" in Chapter 4, where this concept is thoroughly described.
9  See Chapter 4, "Designing the temporal object."
10  Cf. my explanation of Kant's division of the sublime aesthetic experience into three phases: 1) Facing the initiating phenomenon/object; 2) struggling to understand or detect the phenomenon/object; 3) reasoning through the phenomenon/object to help the imagination understand it. See the section on "The stages of the sublime" in Chapter 2.

11  See the section "Turning things upside down" in Chapter 5.
12  See Table 2, Aesthetic strategy model.
13  Cf. the section on "The easily decodable" in Chapter 1.
14  See the section "The easily decodable" in Chapter 1 for a further description of the semiotic terms connotations and anchoring.
15  Cf. the section "A specific example of DSR" in Chapter 6
16  By dogma I mean "rules" that the designer or company invent in order to create direction in the product development.
17  See the section on the sublime in Chapter 2, where I describe the cathartic experience in detail.
18  See the Aesthetic strategy model.
19  Cf. my explanation of Kant's division of the sublime aesthetic experience into three phases in Chapter 2.
20  See the section "Adhering to universal aesthetic principles" in Chapter 1.
21  Cf. the section "The universal effect of color" in Chapter 1.
22  Cf. the section "The need for structure and balance" in Chapter 1.
23  See the section "The experience of minimal inertia" in Chapter 1 for a further elaboration of the term "inertia."
24  See the section "The stages of the sublime" in Chapter 2.
25  Cf. the section "Fleeting beauty" in Chapter 3.
26  See the section "A Specific Example of DSR" in Chapter 6.
27  Cf. the section "When an object assumes its character" in Chapter 3.
28  Cf. the section "Wabi-sabi aesthetics" in Chapter 3.
29  See the section "Fleeting beauty" in Chapter 3.
30  See the sections "The beautiful" and "Adhering to universal aesthetic principles" in Chapter 1.
31  Cf. the section "Zeitgeist analysis" in Chapter 3.
32  Cf. the section "Wabi-sabi aesthetics" in Chapter 3.
33  Cf. the section "The time of becoming" in Chapter 4.
34  Cf. the section "Adhering to universal aesthetic principles" in Chapter 1.
35  See the Aesthetic strategy model.
36  See the section "Zeitgeist analysis" in Chapter 3, in which I describe the Schein analysis model consisting of the artifact level, the espoused value and the basic underlying assumptions.

# Bibliography

Aristotle. 1996. *Poetics*. London: Penguin Books.

Arnheim, Rudolf. 1974. *Art and Visual Perception*. Berkeley: University of California Press.

Barthes, Roland. 1977. *Image-Music-Text*. Edited and translated by Stephen Heath. London: Fontana Press.

—. 2001. *A Lover's Discourse: Fragments*. Translated by Richard Howard. New York: Hill and Wang.

Baudelaire, Charles. 1998. *The Flowers of Evil*. Translated by James McGowan. Oxford: Oxford University Press.

—. 1964. *The Painter of Modern Life and Other Essays*. Translated by Jonathan Mayne. London, UK: Phaidon Press.

Benjamin, Walter. 2007. *Illuminations: Essays and Reflections*. Translated by Harry Zohn. New York: Schocken Books.

Bille, Mikkel and Tim Flohr Sørensen. 2012. *Materialitet: en indføring i kultur, identitet og teknologi* [*Materiality: An Introduction to Culture, Identity and Technology*]. Frederiksberg: Samfundslitteratur.

Böhme, Gernot. 2010. On Beauty. *Nordic Journal of Aesthetics*, 39, pp. 22–33.

Brubach, Holly. 2012. The Right Stuff. *New York Times Style Magazine*, [online] October 4. Available at: http://tmagazine.blogs.nytimes.com/2012/10/04/the-right-stuff-orhan-pamuk/

Brøgger, Stig, Else Marie Bukdahl and Hein Heinsen, eds.1985. *Omkring det Sublime* [*Around the Sublime*]. Copenhagen: The Royal Danish Academy of Fine Arts.

Burke, Edmund. 1958. *A Philosophical Enquiry into the Origin of our Ideas of the Sublime and Beautiful*. London: Routledge and Kegan Paul.

Chapman, Jonathan. 2011. *Emotionally Durable Design*. London: Earthscan.

Crowther, Paul. 1989. *The Kantian Sublime: From Morality to Art*. Oxford: Clarendon Press.

Fletcher, Kate and Lynda Grose. 2012. *Fashion & Sustainability: Design for Change*. London: Laurence King Publishing Ltd.

Gotfredsen, Lise. 1998. *Billedets formsprog* [*The Idiom of the Image*]. Copenhagen: Gads Forlag.

Goethe, J.W. von. 2004. *The Sorrows of Young Werther*. Translated by Burton Pike. New York: Random House.

Guldager, Susanne and Kristine Harper. 2015. *The Empowering Experience of Working With Local Artisans*. Less Magazine, [online] 4, pp. 103–09. Available at: http://lessmagazine.com/issue-04/

Habib, M.A.R. 2005. *A History of Literary Criticism and Theory: From Plato to the Present*. Malden, MA: Blackwell.

Itten, Johannes. 1997. *The Art of Colo.* The Subjective Experience and Objective Rationale of Color. Hoboken, NJ: John Wiley and Sons.

Juniper, Andrew. 2003. *Wabi Sabi: the Japanese art of impermanence*. Clarendon, VT: Tuttle Publishing.

Jørgensen, Dorthe. 2012. Fornemmelsens filosofi: Æstetik, fænomenologi og erfarings-metafysik [The Philosophy of Sensation: Aesthetics, Phenomenology and the Metaphysics of Experience]. In: Ulla Thøgersen and Bjarne Troelsen, eds. *Filosofi og kunst* [*Philosophy and Art*]. Aalborg: Aalborg University Press.

—. 1990. *Nær og Fjern: spor af en erfaringsontologi hos Walter Benjamin* [*Near and Far: Traces of an Ontology of Experience in Walter Benjamin*]. Aarhus: Modtryk.

—. 2001. *Skønhedens Metamorfose: De æstetiske idéers historie* [*The Metamorphosis of Beauty: The History of Aesthetic Ideas*]. Odense: Odense University Press.

—. 2008. *Skønhed: en engel gik forbi* [*Beauty: An Angel Passed by*]. Aarhus: Aarhus University Press.

Kandinsky, Wassily. 2008. *Concerning the Spiritual in Art*. Translated by Michael T.H. Sadler. Auckland, NZ: The Floating Press.

Kant, Immanuel. 2002. *Critique of the Power of Judgment*. Translated by Paul Guyer and Eric Matthews. Cambridge: Cambridge University Press.

—. 1991. *Observations on the Feeling of the Beautiful and the Sublime*. Berkeley: University of California Press.

Knausgaard, Karl Ove. 2012. *My Struggle: Book One*. Translated by Don Bartlett. New York: Archipelago Books.

Koren, Leonard. 2008. *Wabi-Sabi for Artists, Designers, Poets & Philosophers*. Point Reyes, CA: Imperfect Publishing.

Kundera, Milan. 1998. *Identity*. Translated by Linda Asher. London: Faber and Faber.

—. 1996. *Slowness*. Translated by Linda Asher. New York: HarperCollins Publishers

Long, Rose-Carol Washton. 1980. *Kandinsky: The Development of an Abstract Style*. Oxford: Clarendon Press.

Lyotard, Jean-Francois. 1994. *Lessons on the Analytic of the Sublime*. Stanford: Stanford University Press.

—. 1991. *The Inhuman: Reflections on Time*. Cambridge: Polity Press.

Merleau-Ponty, Maurice. 2002. *The World of Perception*. New York: Routledge.

Ørskov, Willy. 1987. *Den åbne skulptur - og udvendighedens æstetik* [*The Open Sculpture - And the Aesthetics of the Outside*] Essays. Copenhagen: Borgen.

—. 1999. *Samlet: Aflæsning af objekter, Objekterne, Den åbne skulptur* [*Detecting Objects and Other Writings*]. Copenhagen: Borgen.

Pamuk, Orhan. 2010. *The Museum of Innocence*. London: Faber and Faber.

Plato. 1997. Greater Hippias. Translated by Paul Woodruff. In: John M. Cooper, ed. *Complete Works*. Indianapolis/Cambridge: Hackett Publishing Company.

Proust, Marcel. 2004. *Swann's Way*. Translated by Lydia Davis. New York: Penguin.

Pugh, David. 1996. *Dialectic of Love: Platonism in Schiller's Aesthetics*. Montreal & Kingston: McGill-Queen's University Press.

Scheff, T.J. 1979. *Catharsis in Healing, Ritual and Drama*. Berkeley: University of California Press.

Schein, Edgar. 2004. *Organizational Culture and Leadership*. San Francisco: Jossey-Bass.

Schiller, Friedrich. 2010. *Aesthetical and Philosophical Essays, vol.1*. London: Forgotten Books.

Thomsen, Søren-Ulrik. 2002. *Det værste og det bedste* [*The Worst and the Best*]. Copenhagen: Gyldendal.

Thyssen, Ole, ed. 2005. *Æstetisk Erfaring: tradition, teori, aktualitet* [*Aesthetic Experience: Tradition, Theory and Topicality*]. Frederiksberg: Samfundslitteratur.
Walker, Stuart. 2007. *Sustainable by Design*. London: Earthscan.

## Webpages:

www.csrkompasset.dk
www.lessmagazine.com
www.localwisdom.info
www.textiletoolbox.com

# Index of terms

# Index of names

Printed in Great Britain
by Amazon